Nature's Enigma:

The Problem of the Polyp in the Letters of Bonnet, Trembley and Réaumur

c. Pronk del. Ad viv. 1744.

J. v. Schley sculp.

Abraham Trembley in his study at Sorgvliet demonstrates the famous experiment of turning the polyp inside out to Count Bentinck's two sons. Note the powder jars filled with polyps that Trembley kept on his window sills. (Vignette from A. Trembley, *Mémoires, pour servir à l'histoire d'un genre de polypes d'eau douce, à bras en forme de cornes* (Leide, 1744))

Nature's Enigma:

The Problem of the Polyp in the Letters of
Bonnet, Trembley and Réaumur

Virginia P. Dawson

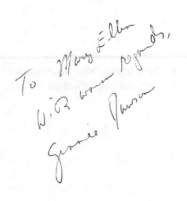

To Mary Ellen
With warm regards,
Ginnie Dawson

American Philosophical Society
Independence Square   Philadelphia
1987

Memoirs of the
American Philosophical Society
Held at Philadelphia
For Promoting Useful Knowledge
Volume 174

Library of Congress Catalog Card No. 86-72886
International Standard Book Number: 0-87169-174-4
US ISSN: 0065-9738

To Dave, Jeff and Emily

CONTENTS

LIST OF ILLUSTRATIONS

Acknowledgments

I have received encouragement and assistance from many people, but first and foremost I owe my gratitude to my husband and children. They have sacrificed, prodded and cheered as my book has taken shape. Without their grace and humor, I never would have finished.

My adviser, Robert Schofield, now of Iowa State University, made me aware of the importance of the cultural setting of science and pointed me in the direction of Geneva. Over the many years that I was his student, he encouraged me not to lose faith that the voluminous and often tedious observations of insects would become the grist for a larger framework of ideas to be discovered. The benefit of his insights I acknowledge with gratitude.

I received the generous assistance of George Trembley of the University of Toronto, who not only opened his private archives at a crucial point in my research, but also shared with me his knowledge of Geneva and the eighteenth century. In addition to the warm hospitality and friendship of George and Nicole Trembley, I enjoyed in Toronto some of the fruits of the scholarship of Maurice Trembley, Professor Trembley's grandfather, who collected many of Bonnet's books, manuscripts and letters.

I would like to thank Philippe Monnier, the curator of the manuscript collection of the Public and University Library of Geneva for his gracious and unflagging help. Years ago, he sent me the microfilm of Réaumur's letters to Bonnet which began my odyssey into eighteenth-century manuscript sources. Later, he provided me with photocopies of Trembley's early letters to Bonnet, as well as the Bonnet-Cramer letters. In addition, he assisted me in the preparation of my catalogue of the Bonnet-Cramer correspondence. I also received the kind assistance of Pierre Berthon, Archivist of the Paris Academy of Sciences, who arranged to have the Bonnet letters to Réaumur microfilmed.

In December 1984 I had the privilege of participating in

a symposium honoring Abraham Trembley, organized by Howard M. Lenhoff of the University of California, Irvine, and by the Société de Physique et d'Histoire Naturelle of Geneva. Through Dr. Lenhoff and his wife, Sylvia, I encountered a delightful group of microbiologists caught up in the study of the modern hydra. Their enthusiastic reading of Trembley, and reexamination of some of his observations through a differently focused lens, enriched my understanding of Trembley's science. The paper that I presented at the symposium, "Trembley's Experiment of Turning the Polyp Inside Out and the Influence of Dutch Science," was published in a volume of proceedings. I am indebted to the Société de Physique et d'Histoire Naturelle for permission to incorporate parts of that essay into the present work.

In addition to the pleasure of meeting members of the Trembley family in Geneva, at that time I had the long awaited opportunity to work in the Public and University Library. There I enjoyed the valuable assistance of Mr. Monnier and his staff in examining other parts of the Bonnet collection.

The scholarship of Jacques Roger and Aram Vartanian contributed immeasurably to my understanding of eighteenth-century biological thought. I also owe a debt to Jean Torlais, Jacques Marx and Bentley Glass for their careful analyses of the work of Réaumur and Bonnet. In addition, I would like to thank Dr. Glass for reading my manuscript and for his helpful comments. Although it does not change my interpretation of Bonnet's early work, I regret that I could not take advantage of the new light shed on Bonnet's mature thought in the recently published *Science against the unbelievers: the correspondence of Bonnet and Needham, 1760–1780* by Renato G. Mazzolini and Shirley A. Roe.

I am grateful to the American Society for Eighteenth-Century Studies for permission to include here some of the research for two articles, "Trembley, Bonnet, and Réaumur and the Issue of Biological Continuity," in *Studies in Eighteenth-Century Culture* 13 (1984):43–63, and "The Problem of Soul in the 'Little Machines' of Réaumur and Charles Bonnet," in *Eighteenth-Century Studies* 18 (1985):503–522. In addition, I have enjoyed and benefited from the lively debates of The Eighteenth-Century Seminar in Cleveland.

I am also indebted to Norma Sue Hanson, Special Collections Librarian, Case Western Reserve University Libraries and Glen Jenkins, Archivist of the Howard Dittrick Museum of Historical Medicine, for their assistance. In addition, plates from Réaumur's *Mémoires pour servir à l'histoire des insectes* (hereafter referred to as *Histoire des insectes*) and Bonnet's *Traité d'insectologie* are reproduced here with the permission of Special Collections of the Case Western Reserve University Libraries.

I received assistance and encouragement from Darwin Stapleton, Patricia Gerstner, Peter Salm, and Alan Rocke, Case Western Reserve University; Georgia Lesh-Laurie and William Shorrock, Cleveland State University; Nelly Hoyt, Smith College; René Taton, University of Paris; Elaine Robson, University of Reading, and Jacques Trembley, Geneva. Patricia Dussaux, in addition to typing my manuscript, added her native ease with French to my own in the laborious deciphering of perplexing eighteenth-century scrawls. Her assistance in transcribing the letters, as well as that of George Trembley, is gratefully acknowledged.

Chapter 1

Introduction

Within a year, Charles Bonnet (1720–93) and Abraham Trembley (1710–84) communicated two striking biological discoveries to the Paris Academy of Sciences. Each had made observations which upset what to the eighteenth-century mind was natural law: animal reproduction required mating, the coupling of the male and female of the same species. Parthenogenesis, proved by Bonnet in July 1740, established that reproduction was possible in aphids without male fertilization. The following March, Trembley brought to the attention of astonished naturalists an even more important discovery. A tiny aquatic animal, the fresh water polyp or hydra, could reproduce like a plant from cuttings. The Parisian community learned of these discoveries through letters to the prominent academician René-Antoine Ferchault de Réaumur (1683–1757), foremost authority on insects in France.

Bonnet's discovery was important in providing the first conclusive evidence of an exception to natural law, but its implications in terms of accepted explanations of animal generation were not as dramatic as the discovery of the regenerative properties of the polyp. Aphid reproduction appeared to give experimental support to the theory of the preexistence of germs, particularly to the view that the procreative vehicle was the egg rather than the sperm.

Although vaguely disquieting to those who confidently lived in a universe held together by God-given natural laws, nevertheless, the discovery of parthenogenesis was not unanticipated. The apparent absence of mating in aphids had been noted by Anton van Leeuwenhoek (1632–1723) and by Giacinto Cestoni (1637–1718). However, while Leeuwenhoek's discovery of spermatozoa supported the idea that the male carried the germ, Cestoni's observations were referred to in the work of an ovist. The French-speaking world was

5

made aware of them through Antonio Vallisnieri's *Oeuvres*, published in Venice in 1733. Both Leeuwenhoek and Cestoni had suggested that aphids might be hermaphroditic, but coming from opposing schools, their intriguing studies of the lowly aphid raised questions about the generation of animals. Experimental proof, however, was lacking.

Réaumur himself performed experiments, reported in the third volume of his celebrated *Mémoires pour servir à l'histoire des insectes* (hereafter referred to as *Histoire des insectes*), to determine whether a female aphid, isolated at birth from contact with other aphids, could reproduce. Though his experiments failed to settle the issue, he pointed out that the only aphids that he had found were female. He had never laid eyes on a male. True hermaphrodites like snails, he observed, in which both male and female were united, still had to mate, but in all the time that he had been observing aphids, he had "never seen any coupling."[1] However, with characteristic restraint when it came to drawing unsupported conclusions, he invited other observers to repeat his experiment. Charles Bonnet built his early reputation on his enthusiastic response to Réaumur's suggestion.

Bonnet's discovery of parthenogenesis was less genius than patience and perseverance. His success in bringing up ninety-five offspring of a single aphid, isolated at birth, observing "not only day by day and hour by hour but even several times in the same hour and always with the glass to render the observations more exact,"[2] earned him the title of *Correspondant* of the Academy. It must have been with a certain satisfaction that Réaumur read Bonnet's letter of 13 July 1740 announcing the discovery to his colleagues in the Paris Academy. He had nearly succeeded in proving parthenogenesis himself. A young man of nineteen had established the fact by using the assiduous observational techniques which he had inspired.

Such was not the case with Trembley's discovery. Réaumur had not anticipated the discovery of the polyp's curious mode

---

[1] René Réaumur, *Mémoires pour servir à l'histoire des insectes* (Paris: Imprimerie Royale, 1735) 3:327.

[2] Bonnet to Réaumur, 13 July 1740, Papers of Réaumur, Bonnet Dossier Biographique (DB), Archives de l'Académie des Sciences, Paris.

of reproduction. Though almost thirty years previously Réaumur had studied the regeneration of the legs of crayfish,[3] there is no indication that this study played a role in the discovery of the polyp. The regeneration of the polyp into complete individuals after it had been cut up was both novel and disturbing. It did not fit easily into the preformationist explanation, nor did it jibe with the accepted concept of the animal machine, a Cartesian legacy. Indeed, rather than lending support to the cause of the proponents of the preexistence theory, the polyp's regenerative properties appeared to give some credence to the theory of epigenesis. The hyperbole of the report published in the *Histoire de l'Académie des Sciences* was a reflection of the excitement and disbelief generated by an "insect" barely more than a quarter of an inch in size which, even after being chopped up, could produce as many new individuals as pieces.

The story of the Phoenix who is reborn from its ashes, fabulous as it is, offers nothing more marvellous than the discovery of which we are going to speak. The chimerical ideas of Palingenesis or regeneration of Plants & Animals, which some Alchemists believed possible by the bringing together and the reunion of their essential parts, only leads to restoring a Plant or an Animal after its destruction; the serpent cut in half, & which is said to be rejoined, gives but one & the same serpent; but here is Nature which goes farther than our chimeras. From one piece of the same animal cut in 2, 3, 4, 10, 20, 30, 40 pieces, & so to speak, chopped up, there are reborn as many complete animals similar to the first.[4]

The report left the reader to draw his own conclusions on

---

[3]René Réaumur, "Sur les diverses reproductions qui se font dans les écrevisses, les omars, les crabes, etc., & entr'autres sur celles de leurs jambes et leurs écailles," *Histoire de l'Académie Royale des Sciences de l'année 1712, avec les mémoires de mathématique & de physique pour la même année* (Paris: Pankoucke, 1777): 295–321.

[4]"L'Histoire du Phoenix qui renait de ses cendres, toute fabuleuse qu'elle est, n'offre rien de plus merveilleux que la découverte dont nous allons parler. Les idées chimeriques de la Palingénésie ou régénération des Plantes & des Animaux, que quelques Alchymistes ont cru possible par l'assemblage & la réunion de leurs parties essentielles, ne tendoient qu'a rétablir une Plante ou un Animal après sa destruction; le Serpent coupé en deux, & qu'on a dit se rejoindre, ne donnoit qu'un seul & même serpent; mais voici la Nature qui va plus loin que nos chimères. De chaque morceau d'un même animal coupé en 2, 3, 4, 10, 20, 30, 40 parties, &, pour ainsi dire, haché, il renait autant d'animaux complets & semblables au premier." *Histoire de l'Académie Royale des Sciences, année 1741* (Amsterdam: Pierre Mortier, 1747) 1:46.

the problem of generation of animals, their similarity to plants, "and perhaps on even higher matters."[5]

This reference to higher matters alludes to the disturbing metaphysical issues which the discovery of the polyp raised. Not only did it seem to present a new mode of animal reproduction that had previously been unsuspected, but it called into question the prevailing mechanistic view of animal biology. Were not animals little machines designed by God according to a preordained pattern? To Trembley's contemporaries, regeneration from parts seemed to introduce an unwelcome element of chance into the regular structure of living beings and implied that regeneration might be an attribute of matter independent of rational plan. Moreover, it brought into focus the problem of animal soul which had been debated by naturalists and metaphysicians since the posthumous publication of Descartes's *Primae cogitationes circa generationem animalium* in 1701.

Recognition of the polyp's amazing properties contributed to a shift in midcentury zoological considerations. The idea that a creature could regenerate without the orderly process of mating, weakened the arguments of those who used the visible world as evidence of God's rational plan. Materialism and vitalism, sometimes associated with pantheistic and atheistic notions, took on a new intellectual respectability. Proponents of the new doctrines attributed the capacity of living matter to organize itself to a force or vital principle, neither reducible to mechanical laws nor directly under God's control. Regeneration implied that matter contained active properties, previously attributed to the benevolence of the Creator. Aram Vartanian was the first to point out that after 1741 questions of function and design embodied in the "biological teleology" of such works as Abbé Noël Pluche's *Spectacle de la nature* (1732) lost ground to materialistic modes of thought.[6] They culminated with La Mettrie's *L'Homme ma-*

---

[5]Ibid., 48.

[6]Aram Vartanian, "Trembley's Polyp, La Mettrie, and Eighteenth-Century French Materialism," *Journal of the History of Ideas* 11 (1950):259–86. See also Shirley Roe's excellent article, "John Turberville Needham and the Generation of Living Organisms," *Isis* 74 (1983):159–84, which illuminates some of the complexities of biological materialism. Elizabeth L. Haigh has examined the vitalism of the Montpellier school in two articles: "The Vital Principle of Paul-Joseph Barthez: The Clash between Monism and Dualism," *Medical History* (Great Britain) 21 (1977):1–14, and

chine and the even more infamous *Système de la nature* of d'Holbach.

In the traditional view accepted prior to the discovery of the polyp, animals were viewed as complex systems of inter-dependent parts; an animal that was cut up was expected to die. That it could reorganize itself and redevelop into a complete animal, perfect in all respects, led to a reexamination of accepted ideas of natural history, a reexamination that had wide ramifications. The newer theories abandoned explanations based on design or purpose. Belief in the providential ordering of the universe by God was no longer necessary.

The turn toward vitalistic and materialistic explanations after 1741 has been documented in the thought of John Turberville Needham, Georges-Louis Leclerc de Buffon, Denis Diderot, Albrecht von Haller, Julien Offray de La Mettrie, Théophile de Bordeu, and Johann Blumenbach, to mention only a few important eighteenth-century biologists and thinkers.[7] The importance of the discovery of the polyp in the shaping of their thought has been pointed out in many separate studies. Yet little attention has been given to the background of ideas which contributed to this shift. For though Trembley's and Bonnet's discoveries were related, and though both received wide publicity in France, neither Bonnet nor Trembley were, in fact, French. Instead they were both members of the tightly knit upper class of the Republic of Geneva and had similar educational backgrounds. An important fact which has been overlooked by historians in assessing Réaumur's influence is that neither Bonnet nor Trembley studied directly under Réaumur. Rather, both received their formal education at the Academy of Calvin, a prestigious institution of higher education in Geneva. Their principal instructors were Gabriel Cramer (1704–52) and Jean-Louis Calandrini (1703–58), two re-

---

"The Roots of the Vitalism of Xavier Bichat," *Bulletin of the History of Medicine* 79 (1975):72–86. In the period after 1741, materialism and vitalism are difficult to separate. It appears that each owed something to the other.

[7]See Jacques Roger, *Les Sciences de la vie dans la pensée française du XVIIIᵉ siècle* (Paris: Armand Colin, 1963) for influence on Needham, Buffon, Diderot and Bordeu; Charles Bodemer, "Regeneration and the Decline of Preformation in Eighteenth-Century Embryology," *Bulletin of the History of Medicine* 38 (1964): 20–31 for influence on Blumenbach; Shirley A. Roe, *Matter, Life and Generation* (New York: Cambridge University Press, 1981) for influence on Haller.

markable mathematicians who helped to shape Genevan science. Trembley and Bonnet were intimate friends, as well as distant relations, actively corresponding with each other as the discovery of the polyp unfolded.

Trembley was the son of a syndic, one of the four highest political officials in the Republic.[8] The youngest of five children, his childhood coincided with a period of social unrest in Geneva. Though little is known of his early education at the Academy, Calandrini was an important influence on it. Trembley wrote his dissertation on the infinitesimal calculus under Calandrini's direction, successfully defending it in 1730.[9]

Apparently Trembley intended to pursue a career as a pastor, but poor health and insufficient means may have influenced his decision to go to the University of Leiden in 1733. He attended for only a short time before finding permanent employment as tutor to the sons of a prominent nobleman, Count William Bentinck. However, Trembley developed connections with the scientific circle at Leiden which included Willem Jacob 'sGravesande (1668–1742), Professor of Mathematics and Philosophy, Hieronymus David Gaubius (1704–80), Professor of Chemistry, and Bernard Siegfried Albinus (1697–1770), Professor of Anatomy and Surgery. Most certainly, he came in contact with the renowned Hermann Boerhaave (1668–1738), Professor of Medicine, as well. He developed a close friendship with Jean-Nicholas Sébastien Allamand (1713–87), tutor to the children of 'sGravesande. Allamand, a fellow Swiss, later became Pro-

---

[8]John R. Baker, *Abraham Trembley of Geneva* (London: Edward Arnold, 1952), is by far the best source for biographical information on Trembley and owes a great deal to the prior unpublished scholarship of Maurice Trembley. Additional biographical information can be found in Paul F. Geisendorf, *Les Trembley de Genève de 1552 à 1846* (Genève: Jullien, 1970). For additional information on Trembley at Leiden see an excellent article by Mouza Raskolnikoff, "De L'Education au Siècle des Lumières: Louis de Beaufort Gouverneur de Prince de Hesse-Hombourg d'après des lettres inédites," *Journal des Savants* (janvier-mars, 1982):77–93. A valuable contribution to the literature on Trembley has also been made by Sylvia and Howard Lenhoff through their translation of Trembley's Memoirs: *Hydra: the Birth of Experimental Biology—1744. Abraham Trembley's Mémoires concerning the Polyps* (Pacific Grove, CA: Boxwood Press, 1986).

[9]Trembley's dissertation, *Theses Mathematicae de Infinitio et Calculo Infinitesimali Quas, Deo Saventi sub Presidio, D.D. Joh. Lud. Calandrini* (Genève: Marci-Michaelis Bousquet, 1730) was exhibited at the Museum of Natural History, Geneva, during the Trembley Symposium, December, 1984.

fessor of Philosophy at Leiden and editor of 'sGravesande's collected works. Pierre Lyonet (1706–89), who resided at the Hague, shared Trembley's enthusiasm for insects. Lyonet, also a correspondent of Réaumur, studied the phenomenon of parthenogenesis, often consulting with Trembley. In turn, through his letters to Bonnet, Trembley shared their observations and perplexities. From 1739 until 1747 Trembley remained on the count's estate at Sorgvliet, about a mile and a half from The Hague. It was there in powder jars teeming with insect specimens that he noticed the peculiar "insect" which Réaumur later named the freshwater polyp.[10]

In contrast to Trembley, Bonnet never left Geneva, except for one trip to Roche to visit his friend von Haller who lived at the other end of Lake Léman. Educated first at home, he was sent to the Academy in 1735 at the age of fifteen. He transferred from the Auditoire des Belles Lettres to the Auditoire de Philosophie the following year to study under Cramer and Calandrini. Though both played a role in his educational development, it was Cramer who became his "oracle."[11]

The importance of this common Genevan background seems to have been overlooked by historians of science. Though some attention has been given to the strong scientific tradition of the Swiss in the eighteenth century, it has focused on the mathematics of the Bernoullis, Leonard Euler and Johann-Heinrich Lambert, and on Albrecht von Haller's physiology. However, in natural history, the French-speaking Swiss of Geneva formed an informal circle around Charles Bonnet and established a tradition in the natural sciences which included Horace-Bénédict de Saussure, Jean Senebier, Pierre and François Huber, Jean-André de Luc, Nicholas de

---

[10]The hydra is not in fact now classified as an insect but as a coelenterate. The particular species that Trembley first worked on is known in modern taxonomy as the *chlorohydra viridissima*. The green hydra was first named *Hydra viridissima* by Pallas in 1766. Thus, this name takes precedence over *Hydra viridis*, given by Linnaeus in 1767. See Paul J. McAuley, "Variation in Green Hydra. A Description of Three Cloned Strains of *Hydra viridissima* Pallas 1766 (Cindaria: Hydrozoa) Isolated from a Single Site," *Biological Journal of the Linnean Society* (1984) 23:1–13. Trembley was familiar with two other species: *Hydra vulgaris* (also named by Pallas) and *Hydra oligactis*.

[11]*Mémoires autobiographiques de Charles Bonnet de Genève*, ed. Raymond Savioz (Paris: Vrin, 1948), 46. Hereafter cited as Charles Bonnet, *Mém. aut.*

Saussure, and Auguste de Candolle.[12] Joined to this natural history circle through correspondence were Albrecht von Haller, Lazzaro Spallanzani and to a lesser extent Jean-Nicholas Sébastien Allamand.[13]

To examine the foundations of this Genevan biological tradition which begins with Bonnet's and Trembley's discoveries, the present study focuses on the correspondence between Trembley, Bonnet, and Réaumur. These letters, many of which are unpublished, as well as the published works of the three scientists, support the view that the discovery of the polyp was substantially due to the independence of the Genevan scientific tradition from that of the French. This conclusion contradicts previous scholarship, which in general has supported Jean Torlais's view that Réaumur played the dominant role in the intellectual formation of both Trembley and Bonnet.[14] This assessment of Réaumur's influence on Bonnet has been repeated by Jacques Marx in *Charles Bonnet contre les lumières*. He calls Bonnet the "intellectual son of Réaumur."[15] More recently, Lorin Anderson has written in *Charles Bonnet and the Order of the Known:*

Considering the encouragement that Bonnet was receiving from Réaumur and the lack of understanding from his father at this time, it seems likely that Réaumur, in the role of mentor reinforcing the inclinations and interests of the son, now began to stand *between* father and son and, in effect, displaced and inherited the father's authority.[16]

---

[12]Of those Genevans mentioned, only Jean André de Luc was not educated at the Academy of Calvin. For an excellent sociological study of Genevan science see Cleopatra Montandon, *Le Développement de la science à Genève aux XVIIIᵉ et XIXᵉ siècles: le cas d'une communauté scientifique* (Vevey: Editions Delta, 1975).

[13]The Allamand correspondence with Trembley is in the George Trembley Archives. The Bonnet-Spallanzani correspondence has been published by Carlo Castellani, ed., *Lettres à M. l'Abbé Spallanzani de Charles Bonnet* (Milano: Episteme Editrice, 1971). See also Otto Sonntag, ed., *The Correspondence between Albrecht von Haller and Charles Bonnet* (Bern: Hans Huber, 1983).

[14]Jean Torlais, "Un Maître et un élève. Réaumur et Charles Bonnet (d'après leur correspondance inédite)," *Gazette Hebdomadaire des Sciences Médicales de Bordeaux* (9 October 1932) 41:641–55 and *Réaumur: un esprit encyclopédique en dehors de l'Encyclopédie* (Paris: Albert Blanchard, repr. 1961), 141–56.

[15]Jacques Marx, *Charles Bonnet contre les lumières: 1738–1850* (Oxford: *Studies on Voltaire and the Eighteenth Century*, 1976): 156–57, 314.

[16]Lorin Anderson, *Charles Bonnet and the Order of the Known* (Boston: Reidel, 1982), 4.

Certainly, one ought not deny the influence on all serious students of insects of Réaumur's six-volume work, *Mémoires pour servir à l'histoire des insectes,* published between 1734 and 1742, or the importance of Réaumur's continued encouragement of Bonnet and Trembley through his letters. Bonnet's assessment of Réaumur's influence is unequivocal: "I found in the *Mémoires* of Mr. de Réaumur all that could satisfy my ardent curiosity, feed my taste, enlighten my Spirit and guide it in its path. I had observed up to that time only by instinct . . . he taught me how to see and rendered me an observer."[17] In 1737, while he was still a student, Bonnet first read the early volumes of Réaumur's *Histoire des insectes.* He then sent Réaumur some of his observations of "liveried caterpillars" in July 1738. Réaumur responded enthusiastically that he was surprised that Bonnet was only a student in philosophy because he appeared "already a master in the art of observing insects."[18] Bonnet then used the copious notes of his observations of insects to compose three discourses on insects which he delivered, probably as lectures, to students and professors comprising the Auditoire de Philosophie at the Geneva Academy.

Trembley's association with Réaumur began after he had left Geneva. In 1739 he read the early volumes of Réaumur's *Histoire des insectes.* In July 1740 he wrote to Bonnet of his discovery of two species of insects which feed on woolen materials and which he thought might be unknown to Réaumur. Speaking of his deep respect for Réaumur as an observer, he wrote: "Above all I am disposed to follow Réaumur and I certainly have had the opportunity to admire him . . . If Réaumur does not know of them, it must be that there are not any in France."[19] Following Bonnet's suggestion, Trembley sent these observations to Réaumur in September 1740. To his embarrassment, they had already been described as "false moths" by Réaumur in his *Histoire des insectes.* Réaumur's tactful response began a correspondence which

---

[17]Charles Bonnet, *Mém. aut.,* 51.

[18]Réaumur to Bonnet, 22 July 1738, Ms Bonnet 42, Bibliothèque Publique et Universitaire de Genève.

[19]Trembley to Bonnet, 26 July 1740, Ms Bonnet 24, Bibliothèque Publique et Universitaire de Genève.

lasted seventeen years and included almost two hundred letters.[20]

There is no doubt that Réaumur was for both Bonnet and Trembley the model and inspiration for their own meticulous observations of insects. However, observation alone does not lead to discovery. To "discover" something by observation a scientist must have an idea of what is there to be discovered. Preconceptions derived from training, background, and experience are incorporated into the design of an experiment, and scientists from different cultural backgrounds are likely to have different interpretations of the same facts. While the Genevans shared a common language, they were Protestant, separated from France both politically and socially. The difference between the French and Genevan approaches to biological questions emerges clearly through the letters of Bonnet, Trembley and Réaumur.

To call either Bonnet or Trembley Réaumur's "intellectual son" is to overstate Réaumur's role in their development. The letters to Réaumur reveal an enormous amount of empirical detail, which Réaumur required and expected, but few ideas. Though observations are meticulously described, there are few instances in which the writers reveal their interpretations of what these observations meant. This they reserved for their letters to each other, and in the case of Bonnet, his letters to his teacher at the Academy of Geneva, Gabriel Cramer. Moreover, when Bonnet later wrote that it was the study of insects which led him to metaphysics,[21] it was not to his own discovery of parthenogenesis, which he owed at least in part to Réaumur, that he was referring, but rather to his study of the regeneration of worms, inspired by Trembley's discovery of the polyp. The polyp's ability to reproduce by sectioning unsettled him intellectually; he wrote that it "upset all my ideas and put my mind, so to speak, in combustion."[22]

One of the values of looking at Bonnet and Trembley

[20]With the exception of three letters in the Archives of the Paris Academy of Sciences, dated 27 August 1744, 22 October 1744 and 18 February 1745, all of the known letters between Réaumur and Trembley have been published by Maurice Trembley, ed., *Correspondance inédite entre Réaumur et Abraham Trembley* (Genève: Georg, 1943). The originals of Réaumur's letters are in the George Trembley Archives.

[21]Charles Bonnet, *Mém. aut.*, 81.

[22]Ibid., 65.

together and assessing their work in terms of a Genevan tradition is the light that it seems to throw on Bonnet's later penchant for metaphysical speculation. The philosophical works of Bonnet published after 1760, *Considérations sur les corps organisés*, *La Palingénesie philosophique*, and *Contemplation de la nature*, were not written in an intellectual vacuum produced by the failure of Bonnet's eyesight in 1747. Yet Bonnet's early work on insects has been regarded as strictly empirical. For example, Elizabeth Gasking has written: "Throughout this period Bonnet showed little interest in theoretical questions and none in the problem of generation. His work is remarkable for its Baconian approach. —A great number of experiments are described in detail, but few conclusions are drawn."[23] Indeed, Bonnet's *Traité d'insectologie*, his first publication which appeared in 1745, seems devoid of metaphysics, except for his discussion of a chain of being based on the polyp as the intermediate form connecting the animal and vegetable kingdoms, which appears in the introduction.[24] Bonnet's discussion here of what he calls a "Ladder of natural Beings" seems anomalous because it is appended to a work bursting with the minutiae of his early observations. These include caterpillars, worms, and aphids, and there is no attempt in the text to relate the empirical facts to the grand scheme outlined in the introduction.

This discontinuity in the thought of Bonnet between the early empirical observations of insects and his later "speculations" is less dramatic than it appears when only published sources are used. Bonnet's early letters to Trembley and Réaumur reveal that even in this early period he was concerned about the issues of animal soul, the chain of being, and the preexistence of germs.

---

[23]Elizabeth Gasking, *Investigations into Generation, 1651–1828* (Baltimore: Johns Hopkins Press, [1967]), 117. F.J. Cole, *Early Theories of Sexual Generation* (Oxford: Clarendon Press, 1930), 199, has called Bonnet's later philosophical works "perverse" and observes with regret that "a more disciplined and sagacious development of his great powers might have made him the greatest naturalist of his time."

[24]Lorin Anderson has discussed Bonnet's concept of the chain of being in *Charles Bonnet and the Order of the Known*, chapter 2. His discussion of Bonnet's intellectual development in chapter 1 is an exception to the existing literature on Bonnet in recognizing the importance of Bonnet's work on insects in leading him to metaphysics, though Anderson does not go further than to cite Bonnet's *Mémoires autobiographiques* as his source for this view.

Trembley also has been branded with the "Baconian" label. He has been presented by historians as one of the great empiricists whose lucky discovery of the polyp was coupled with exacting observational technique. Philip Ritterbush in his *Overtures to Biology* has written:

Trembley's careful experiments are a measure of the difficulties faced by a sound experimentalist owing to contemporary confusion about the difference between plants and animals. He was indifferent to the idea of botanical analogy and sought clear proof that the polyp was either plant or animal in nature, unlike the later speculators who tried to make it out to be a little of each.[25]

By looking at Trembley and Bonnet together, it is possible to distinguish a common background of ideas which gives meaning to their apparently blind empiricism.

In contrast to the rather one-sided presentation of Bonnet and Trembley by American and British historians of biology, Réaumur has been more perceptively understood by the French historian of science, Jacques Roger. Réaumur's work on insects, coming at the end of the period between 1670 and 1745 which Roger has called that of the "new scientific spirit," exemplifies the tension between the rational and mechanical on the one hand, and the observational and experimental on the other.[26] Réaumur's legacy was Cartesian, but a Cartesianism tempered by a skeptical attitude toward abstract systems. Though early eighteenth-century French science retained a Cartesian emphasis on clear thinking and rational demonstration of facts, which to an extent it has never lost, the Cartesian philosophy in its totality—the belief in a world machine created by God according to immutable laws of matter and motion discoverable through the use of human reason—came under scrutiny. Roger argues that it was precisely this new emphasis on observation and experiment by which Cartesian rationalism was progressively undermined. Observation and experiment were to an extent logically incompatible with the idea of mechanism, a presup-

---

[25]Philip C. Ritterbush, *Overtures to Biology* (New Haven: Yale University Press, 1964), 124.

[26]Jacques Roger, *Les Sciences de la vie dans la pensée française du XVIIIe siècle.* Roger's entire discussion in his chapter, "Le Nouvel Esprit scientifique," 163–254, has been particularly helpful in my discussion of the problem of mechanism.

position unprovable by strictly empirical means. Particularly as applied to biology, the concept of the animal machine, the idea that the internal structure and configuration of organs in the living organism could be viewed as merely a rather complex system of levers, pullies, gears and hydraulic pumps, was vulnerable as knowledge increased. Roger has eloquently described the increasing incompatibility of the mechanical idea with the new facts revealed by observation: "Born of an imperious need for clarity, very satisfying for the intelligence who sees things from above, a mechanical conception of life is destined to be buffeted at every moment by some facts revealed by observation and for which it cannot account."[27]

The invention of the microscope at the beginning of the seventeenth century and its skillful use by such early practitioners as Anton van Leeuwenhoek, Marcello Malpighi, and Jan Swammerdam greatly assisted the empirical cause at the expense of the internal consistency of the Cartesian system. The extraordinary vogue of the new instrument on a popular as well as a scholarly level succeeded in turning the dominant emphasis of biological study away from human anatomy towards a preoccupation with the serious study of insects. This study tended to have unsettling metaphysical implications. With the unveiling of the complex structures in insects, the superiority of man as the object in creation to which all other animals were subordinate was called into question. Were not insects even more admirable than man because of their complexity, their minuteness, the skillful artistry which their study revealed? "Insects threw the scholars out of their ruts, refused to be placed in traditional frames of reference, ruined the most solid analogies and the most accepted laws."[28]

In the Cartesian system the belief in the animal machine was supported by the evidence of a rationally ordered universe which operated according to natural law. A corollary to this belief in uniform laws was the reliance on the argument from analogy. All animals were believed to have analogous structures. The study of the internal anatomy of insects revealed heart, viscera and reproductive organs analogous to the same structures in larger animals. Comparable structures

---

[27]Jacques Roger, *Les Sciences de la vie,* 164.
[28]Ibid., 238.

were found in plants. Even animals which were viviparous could be accommodated to the idea of analogy, since their eggs were produced internally.

Asexual reproduction in aphids through parthenogenesis and in the polyp by regeneration after sectioning or by budding was a challenge to the fixed regularity of the Cartesian animal machine. These discoveries, particularly that of the polyp, upset the logically consistent framework which Cartesianism had represented to the early eighteenth-century mind. Trembley's careful study of the polyp, which will be examined in detail in Chapter Four of this study, did not reveal the analogous structures to be expected in the body of an animal. Where were the viscera, the heart, the ovaries or testes? Unlike Bonnet and Trembley, at first Réaumur had difficulty in accepting the exceptions to the general rules that seemed to govern animal reproduction. Since his approach to natural history was Cartesian, he never considered the polyp as an intermediate form, sharing ties with both the vegetable and the animal realms.

The contrast between Réaumur's interpretation of the discovery of the polyp and that of Bonnet and Trembley which this study seeks to establish brings into focus the importance of using both published works and correspondence as sources in the study of eighteenth-century science. Correspondence for the naturalists of the period was a means of official communication of new discoveries. This was particularly important for those observers with serious scientific aspirations who lived outside the cosmopolitan centers which had science academies: Paris, London, St. Petersburg and Berlin. Especially for French-speaking foreigners, correspondence with the Paris Academy was a means of establishing scientific priority and reputation. The title of *Correspondant* was officially conferred by the secretary of the Academy if a particular individual's contributions to science were deemed worthy. A select few had the privilege of publishing in *Mémoires de mathématiques et de physique . . .par divers sçavans,* a collection reserved for distinguished foreign contributors. An actual seat in the Academy as an *Associé étranger* carried enormous prestige, though it was practically impossible to earn this distinction, since there were only eight such seats, and the occupants were elected for life.

Correspondents were assigned to regular members of the Academy, and there was no limit on the number of correspondents a particular academician could enjoy. Réaumur, a sweet-tempered bachelor whose only diversion from his scientific research and arduous official duties at the Academy was weekly attendance at the salon of Madame de Tencin, devoted an extraordinary amount of energy to his correspondence. Trembley and Bonnet were among his approximately one hundred correspondents whose letters were sent through the Count Pajot d'Onsenbray, Director of the Royal Postal Services, with an inner envelope addressed to Réaumur. This represented a considerable saving in postal fees for Réaumur's correspondents, since he expected any serious natural history descriptions to be accompanied by actual specimens and he took pains to describe in his letters how these specimens ought to be prepared for shipping. Réaumur's correspondents included all those who were serious investigators of insects in the early part of the century—not only Trembley and Bonnet, but also Charles de Geer (1720–78) in Sweden, Gilles Auguste Bazin (1681–1754) in Strasbourg, and Pierre Lyonet in Holland.[29] Réaumur read the contents of letters he considered significant at regular meetings of the Academy and such communications were recorded in the *Procès-verbaux*.

The letters of Bonnet and Trembley to Réaumur are examples of official correspondence. They knew that their letters might be read to the assembled Academy, and for this reason they are painstakingly composed. As the drafts of the letters written by Trembley to Réaumur show, Trembley worked over his letters, making many corrections and deletions.[30] Bonnet described his letters to Réaumur as small treatises and complained about the amount of time their com-

---

[29]A complete list of Réaumur's correspondents can be found in Jean Torlais, "Inventaire de la correspondance et des papiers de Réaumur conservés aux Archives de l'Académie des Sciences de Paris," in *La Vie et l'oeuvre de Réaumur* (Paris: Presses Universitaires de France, 1962), 13–24.

[30]The drafts of these letters can be found in the George Trembley Archives, Toronto, Ontario.

position required.[31] Their length was excessive. It was not
unusual for Bonnet to write a letter of fifteen to twenty pages
of minuscule handwriting. His enthusiasm to demonstrate
his competence as an observer to the esteemed academician
prevented him from leaving out the least detail. Though at
the time of his investigation of reproduction in aphids, or
plant lice, he was but a nineteen-year-old student at the Academy of Geneva, he viewed his formal studies as a tedious
diversion from his work on insects. In his first letter to Réaumur he claimed that "the sweetest moments, those in which
I have enjoyed the pure pleasures of an innocent pastime,
have been spent near my Insects."[32] None of these rapturous
moments appear to have gone unrecorded.

In contrast, Réaumur dashed off his responses, seldom
bothering with punctuation. He was at the height of his career, an extraordinarily busy man, involved not only in research and publication of his *Histoire des insectes*, but also at
the time of his first letters to Bonnet in 1738 working on
comparative thermometric observations and studies on the
evaporation of snow. The following year he became assistant
director of the Academy of Sciences and continued his experiments on new ways to make porcelain. In 1740, the year
he began corresponding with Trembley, he became director
of the Academy, an honor which he received nine times. It
is no wonder that his early letters to Bonnet and Trembley
were relatively short, though as their work became more
substantial, his replies lengthened. Since Trembley and Bonnet were composing official letters, they were respectful and
restrained. Réaumur was a venerable and scientist of repute.
They were young and unknown.

The letters which Bonnet and Trembley exhanged with
each other, however, have a completely different ring. They
are informal and often intimate; they provide insight into
how their ideas differed from those of Réaumur. Bonnet

---

[31]Bonnet to Trembley, 24 March 1741, George Trembley Archives, Toronto,
Ontario.
[32]"Je puis vous l'asseurer Monsieur, depuis que je me suis procuré la lecture de
vos Livres, depuis que je leur ai donné toute mon application, les plus doux moments, ceux ou j'ai joui des plaisirs purs d'un Amusement innocent, je les ai passé
auprès de mes Insectes." Bonnet to Réaumur, 4 July 1738, Réaumur (DB), Archives
de l'Académie des Sciences, Paris.

often complained to Trembley that his letters were too "laconic" and he begged him for more details.[33] But at other times he seemed to derive a certain amount of satisfaction from the fact that their communication was sketchy. Though lacking a complete description of Trembley's experiments, he was able to turn up strikingly similar conclusions.[34] Both were delighted at the promise that their new epistolary commerce held. Trembley wrote to Bonnet that his discovery of his young cousin was more precious to him than the discovery of "the rarest insect."[35] The potential for a deepening friendship with his cousin, it seems, far outweighed any discovery in entomology. Bonnet showed equal enthusiasm for this exchange. Good observers were few, he lamented; he would make sure that he would furnish his quota of observations and "neglect nothing in order to make our correspondence equally amusing and instructive."[36]

For the historian, the study of this three-way correspondence among Réaumur, Bonnet, and Trembley, which offers examples of both official and personal correspondence, reveals the value of letters in illuminating the history of thought. Bonnet himself alluded to this function of correspondence in one of his letters to Trembley. Speaking of his correspondence with the Italian naturalist Lazzaro Spallanzani (1729–99), he wrote that letters were to be preferred to finished treatises for they "show better the march of the Spirit in the Search of Truth," and he added: "The letters of the great Leibnitz and those of Clarke have enriched us more than their books: they are in shirt sleeves in their Letters, and often rather buttoned up in their Books."[37] This unbuttoned aspect of personal correspondence allows us to follow

---

[33]Bonnet to Trembley, 27 December 1742, George Trembley Archives, Toronto, Ontario.

[34]Bonnet to Trembley, 28 June 1742, George Trembley Archives, Toronto, Ontario.

[35]Trembley to Bonnet, 4 October 1740, Ms Bonnet 24, Bibliothèque Publique et Universitaire de Genève. Bonnet and Trembley were related through Jean-Antoine Lullin (b.1627), Bonnet's maternal great grandfather, Trembley's maternal grandfather, as noted by W. H. van Seters, *Pierre Lyonet, 1706–1789* (La Haye: Martinus Nijhoff, 1962), 216, note 11.

[36]Bonnet to Trembley, 18 December 1740, George Trembley Archives, Toronto, Ontario.

[37]Bonnet to Trembley, 21 October 1768, George Trembley Archives, Toronto, Ontario.

the process of discovery and, if we are lucky, occasionally catch the glint of philosophical issues among the records of empirical observations. Though twentieth-century biologists continue to be astounded by the detail and accuracy of observations of insects made in the eighteenth century, it is the matrix of ideas from which these observations emerged that is of interest to the historian.[38] They are the elusive metal assayed from the dross of empirical detail. Particularly with regard to the discovery of the polyp, the issue of animal soul emerges as a metaphysical impasse to which the observations themselves offered no solution.

The letters of the early years of the 1740s chronicle the unfolding of the discovery of the polyp and comprise Bonnet's formative period. It is for this reason that the letters which Bonnet and Trembley exhanged are important. Through them, common threads that can be related to their Genevan background are apparent. After 1747 their lives took radically different paths, though their correspondence continued until Trembley's death. Bonnet became increasingly interested in metaphysical problems, for reasons of health seldom leaving his home in Genthod except for short excursions into Geneva. His extensive correspondence became a substitute for direct interaction with the intellects of the Republic of Letters. In contrast, Trembley, upon leaving the household of Count Bentinck, in 1750 became the personal tutor of Charles, the Duke of Richmond. This employment included extensive travel, through which Trembley maintained extensive scientific contacts, but which left little time for scientific research. Thus both men, in different circumstances and presumably for different reasons, turned away from their early passion for insect studies.

Trembley and Bonnet emerged from a unique tradition of science. Though rational in spirit, Genevan science did not represent a simple importation of the Cartesian system. At first bitterly resisted because of its association with ideas

---

[38]For example, Réaumur's *The Natural History of Ants*, translated and annotated by William Morton Wheeler (New York: Knopf, 1926); *Histoire des scarabées. Encyclopédie entomologique*, edited by P. Lesné and F. Picard (Paris: Paul Lechevalier, 1955) and I.I. Kanaev's *Hydra: Essays on the Biology of Fresh Water Polyps* were published by scientists, inspired by the unrecognized excellence of eighteenth-century work on invertebrates.

that were regarded as heretical, Cartesianism took root at the same time that the old orthodox Calvinism was adjusted to reflect newer, more liberal ideas of God and predestination. This study seeks to establish the importance of the education that Trembley and Bonnet received at the Academy of Calvin in Geneva and illustrates the influence of Dutch Newtonian science through 'sGravesande and Boerhaave, and Lcibnizian ideas, possibly derived from contact with other Swiss scientists. The letters that Trembley, Bonnet, and Réaumur exchanged reveal that Trembley and Bonnet did not conceptualize the problems that the discovery of the polyp engendered in exactly the same way as Réaumur did. The Bonnet-Trembley correspondence has received particular attention here because none of the letters has been published in full.[39] Because Bonnet's letters to Trembley are still in the private archives of Professor George Trembley, they have not received any attention by scholars, except for the publication by Maurice Trembley of brief quotations in the notes of *Correspondance inédite entre Réaumur et Abraham Trembley*. Their study reveals the common ground that Trembley and Bonnet shared. Finally, a consideration of the letters which Bonnet exchanged with his teacher at the Academy, Gabriel Cramer, reinforces this view of the unique character of Genevan science. By considering the polyp as more than the product of skillful observation and by attempting to situate the discovery and Bonnet's and Réaumur's reactions to it within a context of ideas, the complicated and fascinating texture of eighteenth-century biology is better revealed.

---

[39]Maurice Trembley is the only scholar who has studied these letters. The series of lectures which he gave in Geneva on his studies was never published. I have included in Appendix A the texts of the Bonnet-Trembley letters to 1744, the date of the publication of Trembley's work on the polyp. George Trembley and I hope to collaborate on an edition of the entire correspondence.

Chapter 2

The Ragged Cartesian Fabric of Eighteenth-Century Biology

Vignette which introduces Réaumur's *Mémoires pour servir à l'histoire des insectes,* vol. III, reflects the aristocratic enthusiasm for insects, in this case the ephemeral mayfly emerging from the aquatic nymph. (Permission to reproduce from Réaumur's *Mémoires,* courtesy of the Rare Book Collection of The Cleveland Health Sciences Library, Historical Division.)

To understand and assess the contributions of Bonnet and Trembley to eighteenth-century theories of animal generation and to understand how they differed from Réaumur, it is necessary to consider the intellectual fabric of eighteenth-century biology, particularly as it relates to the general background of French Cartesianism. The apparent similarities between the three naturalists has understandably led historians to overestimate Réaumur's influence. Since Trembley, Bonnet, and Réaumur all refer to the insects which are the objects of their studies as "little machines," on the surface it might seem that each was similarly impressed by the Cartesian image of the animal machine. The idea of a clock-like universe governed by natural law, popularized by Bernard de Fontenelle in his *Entretiens sur la pluralité des mondes,* dom-

25

inated early eighteenth-century thought. Education in Geneva, as well as in Paris, was Cartesian in spirit. For example, Bonnet wrote in his *Mémoires autobiographiques:* "I harbored at least the germ of the geometrical spirit which Fontenelle said was more precious than geometry itself. The spirit of observation and analysis, which would one day develop in me, does not differ from the geometrical spirit."[1] However, Descartes's monolithic achievement of a logically consistent deductive system could not have been expected to have survived intact nearly a century after it was conceived.

While Réaumur can be considered more faithful to Cartesian philosophy than his Genevan counterparts, even he had strong reservations with regard to Descartes's extravagant claims that natural laws were in the power of man to comprehend. His Cartesianism had been modified by a study of Malebranche. He concluded from his observations of insects that the Cartesian faith in the transparence of the universe to human reason was an illusion. Application of the microscope to the study of the various structures within the bodies of insects revealed mechanism upon mechanism of marvelous artistry produced by God, only a portion of which was accessible to the human investigator. Réaumur consistently discussed the "particular mechanisms" of insects, but he strictly avoided general statements about the mechanical laws undergirding the entire system.[2] The mechanical complexity involved in the design of an insect was worthy of arduous study because it attracted the admiration of the investigator. The totality of the Supreme Artisan's Creation was unfathomable.

Like Malebranche, Réaumur stressed the limits of human knowledge, which he believed "did not reach beyond the first crust of several of the small particles of the universe."[3] Each being in the universe contributes to the perfection of the whole, but "how can we have the least idea of the infinity and the necessity of these combinations when we do not know

---

[1]Charles Bonnet, *Mém. aut.*, 48.
[2]Jacques Roger, *Les Sciences de la vie dans la pensée française du XVIII*ᵉ *siècle*, 224.
[3]Réaumur, *Histoire des insectes* 5:xliii. Quoted by Jean Torlais, "Réaumur philosophe" in *La Vie et l'oeuvre de Réaumur* (Paris: Presses Universitaires de France, 1962), 150.

the ones which must enter into a grain of common earth!"[4] The earth itself may only be a particle in the immensity of the universe.[5]

On the subject of final causes Réaumur, unlike Bonnet, retained a Cartesian distrust of teleological explanations. Though he accepted the view that the scientist could study the particular mechanisms that revealed the artistry with which each insect was constructed, he denied that it was always possible or desirable to fathom the purposes for which they were designed. The purpose of a careful study of the structure of insects and their behavior was the better to admire the wisdom and perfection of the Creator; to speak of final causes was a presumption of human inquiry. It should be enough to judge the truth or falsehood of the observed fact. The designs or intentions of God were impenetrable.

Réaumur specifically made a plea for prudence and restraint when discussing the ends for which a particular mechanism was designed. "A desire which cannot be praised enough," he wrote,

that of attributing great ideas to the author of the universe, the better to see the extent of his providence, has led those who have wished to assign final causes to facts and observations which insects have furnished them to many judgments which were too precipitous and to many false conclusions.[6]

Réaumur asked how we could claim that wings were made for the purpose of flying, for example, when not all insects which have wings can fly?

All that we can conclude is that we must be extremely circumspect in the explanation of ends used by one whose secrets are impenetrable; we often praise badly a wisdom which is so greatly above our praises. Describe with the greatest precision possible his products; that is the manner of giving praise which is most suitable for us.[7]

Réaumur expressed his dismay when, after the failure of Bonnet's eyes in 1747, Bonnet seemed to be straying into

---

[4]Ibid.
[5]Ibid.
[6]Réaumur, *Histoire des insectes* 1:23. Quoted by Jacques Roger, *Les Sciences de la vie*, 248.
[7]Réaumur, *Histoire des insectes* 1:25. Quoted by Jacques Roger, *Sciences*, 248.

teleological explanations without testing them by actual observations. "You are too wise," he wrote, "to permit yourself to always guess the reasons for varieties and final causes. You are forming only suspicions which can provide curious experiments appropriate for confirming your suspicions or destroying them."[8]

Though in general Réaumur rejected explanations in terms of final causes, he did on occasion make qualified and careful suggestions that a certain fact might indicate a conclusion with regard to design or purpose. However, the observed fact is always the focus of his concern, not the speculations that a certain fact might suggest. For example, in discussing the multiplication of worms by cuttings, Bonnet asked Réaumur in a letter written 5 November 1741 the reason why God had accorded the property of regeneration to some insects. "But for what end could this Wisdom, which does nothing in vain, intend in giving to Insects a property which the Animals which we judge the most excellent, Man even, cannot keep himself from envying?"[9] In answering Bonnet's question in his response, Réaumur made only a qualified suggestion that animals which were only partly eaten by others were designed with the ability to reproduce the part that had been eaten.

If you were to wish to guess the ends of nature, you might suspect that the animals which must serve as abundant food for others, but which are ordinarily only eaten in part, have in the remaining part something with which to reproduce the part that has been eaten . . . the animals whose bodies break very easily also need this source of reproduction which has been granted to them as well as to crayfish with respect to their legs.[10]

From the beginning of their correspondence, Bonnet is interested in the purposive activity of insects. For Bonnet, the assiduous observation of insects revealed evidence of God's design. It was through Abbé Pluche that Bonnet came

[8]Réaumur to Bonnet, 5 June 1747, Ms Bonnet 26, Bibliothèque Publique et Universitaire de Genève. See also Jacques Marx, *Charles Bonnet contre les lumières: 1738–1850*, 319.

[9]Bonnet to Réaumur, 4 November 1741, Papers of Réaumur, Bonnet (DB), Archives de l'Académie des Sciences, Paris.

[10]Réaumur to Bonnet, 30 November 1741, Ms Bonnet 26, Bibliothèque Publique et Universitaire de Genève.

to Réaumur. In fact, when Bonnet recounts in his *Mémoires autobiographiques* his first encounter with Réaumur's work, it is linked in his mind with that of the *Spectacle de la nature,* which gave him his first taste for the study of insects. Arriving one day at the home of Ami de la Rive (1692–1763), his professor of logic, he chanced to see a copy of Volume One of Réaumur's *Histoire des insectes* on a table. "A mechanical movement made me open it; I fell by chance on some illustrations which showed different species of caterpillars in their natural attitudes. I felt myself extremely moved."[11] Bonnet attempted to borrow it from his professor, but he was rebuffed with the advice that he should read Pluche instead. Bonnet revealed that not only had he read Pluche, but he knew the work by heart.[12] It was his encounter with the *Spectacle de la nature* at the age of sixteen that had been instrumental in directing his youthful interests to natural history. The book had fallen by chance into his hands. Opening it to the description of the ant-lion, Bonnet immediately felt "a sensation which I can only compare to that which Malebranche experienced at the reading of *L'Homme* of Descartes. I did not read the book, I devoured it. It seemed to me that he developed in me a new sense or new faculties; and I would have willingly said that I was only beginning to live."[13] Bonnet immediately began his own observations of the ant-lion. Their exactitude carried him well beyond those of François Poupart, the authority which Pluche had used.

Abbé Pluche's *Spectacle de la nature* was an extreme example of a work that used the argument from design to encourage the pursuit of natural history. However, Pluche did not regard the value of natural history as the dispassionate acquisition of fact, but as the means to instill moral virtue in the young. By the early eighteenth century there were many works which argued that the natural world demonstrated the perfection of God. The popularity of Pluche shows how far from Descartes's strict avoidance of teleological explanations the practice of natural history had moved. Insects were considered the best examples of biological teleology, since each

---

[11]Charles Bonnet, *Mém. aut.,* 48.
[12]Ibid.
[13]Ibid., 42.

tiny structure seemed so complex, so perfect, that it could
not have been the product of chance.

The increasing popularity of the microscope and the decline of Cartesian rationalism coincided with a movement of
ideas called insecto-theology.[14] The design of the wing of a
butterfly, the life cycle of a mayfly, or the remarkable ad-
aptations of the ant-lion were seen as examples of God's
artistry. There were many works from the late seventeenth
century which argued that the Book of Nature demonstrated
the perfection of God in the same way that the Bible demonstrated revelation. An emphasis on empirical investigations
was accompanied by an increasingly literal interpretation of
the Bible. These seemingly irreconcilable sources of divine
knowledge can be seen particularly in the work of Swammerdam. Friedrich Christian Lesser's work, *The Theology of
Insects, or Demonstration of the Perfections of God in all that concerns Insects*, published in German in 1738 and translated into
French in 1742 by Pierre Lyonet, is an example of a genre
of teleological science which drew inspiration from William
Derham's *Physico-Theology: Or, a Demonstration of the Being and
Attributes of God from His Works of Creation* (1704). It is possible
that a number of important studies of insects in the eighteenth century are related to insecto-theology, though all have
been regarded as examples of rigorous empirical studies.
Bonnet's *Traité d'insectologie* (1745), Trembley's *Mémoires, pour
servir à l'histoire d'un genre de polypes à bras en forme de cornes*
(1744), Gilles-Auguste Bazin's *Histoire naturelle des abeilles*
(1744), Pierre Lyonet's *Traité anatomique de la chenille qui ronge
le bois de saule* (1760), and Charles de Geer's *Mémoires pour
servir à l'histoire des insectes* (1752–78) may be related to this
tradition, though no definitive analysis of the theology undergirding these works has been made. In general, it seems
that the Protestant investigators were more enamored of the
design argument than were their Catholic counterparts,
though further research will be necessary fully to establish
this point.

While Réaumur and Bonnet differed in their degree of

---

[14]Very little has been written on insecto-theology. See Francis J. Cole, "Jan Swammerdam, 1637–80," *Nature* (1937), 220 and Vartanian, "Trembley's Polyp," *Journal
of the History of Ideas* 11 (1950):259–86 for brief discussions.

acceptance of teleological explanations, on two fundamental points both agreed in their rejection of Cartesian biological explanations. The first was Descartes's belief that the generation of living beings could be explained mechanically. Descartes claimed that, through jostling and fermentation, particles of seminal fluid, produced by the blood of both male and female, came together in the womb. From this union, the embryo developed epigenetically. This theory found in *De la formation de l'animal,* was published posthumously by Claude Clerselier in 1664. It found few adherents because it failed to explain the reason why development followed consistent patterns, reflected in the fixity of the species. The fact that the union of male and female of the same species always produced similar offspring did not appear to be the product of chance epigenetic development. Réaumur's opinion that perhaps Descartes himself was not satisfied with his account of generation was probably typical of most eighteenth-century thinkers:

The great Descartes did not presume so much upon the strength of his genius, when he tried to explain the formation of the universe, as he did when he attempted to explain the formation of man; nor was he, perhaps, himself over and above satisfied with his essay on this last subject, which was not printed until after his death.[15]

The second point on which Cartesian theory was disputed was the view of animals as soulless automata, subject only to the laws of matter and motion. Descartes had defined soul as rational. To have a soul meant to possess intelligence; therefore, it was reserved for man alone. The view of animals as machines reinforced the basic separation of extended matter and immaterial non-extended soul—the foundation of the Cartesian dualism on which the entire philosophy was based. For Descartes the perfect regularity and precison exhibited by animal behavior demonstrated that animals were like machines devoid of a rational principle. Drawing the analogy between the movements of animals and the mechanical motions of a clock governed by springs, counter-

---

[15]Réaumur, *Art de faire éclore et d'élever en toute saison des oiseaux domestiques de toutes espèces* 2 (Paris, 1749), tr. A. Trembley, 1750. Quoted by Elizabeth Gasking, *Investigations into Generation, 1651–1828,* 68.

weights, and wheels, Descartes believed that life was the result
of the arrangement of parts in the animal that caused it to
function. He wrote:

I desire, I say, that you should consider that all these functions
follow quite naturally solely from the arrangement of its parts neither more nor less than do the movements of a clock or other
automaton from that of its counter weights and wheels; so that it
is not necessary to conceive in it any other principle of movement
or of life than blood and spirits set in motion by the heat of the
fire that burns continually in its heart and is of no other nature
than all those fires that are in inanimate bodies.[16]

According to Descartes, if animals were capable of thought
or intelligence, they would have to be granted an immortal
soul as well. While some of the activities of the higher animals
appeared rational, who could conceive of a sponge or an
oyster with a soul?

An influential early condemnation of the Cartesian theory
on both points was that of Claude Perrault in his *De la mé-
chanique des animaux* (1680). Though he accepted the idea of
animals having a machine-like design, he argued that animals
have souls and he found the Cartesian description of generation both impious and implausible.[17] Common sense
seemed to render the Cartesian theory of generation unlikely,
for, as Fontenelle later quipped, you could put a she-dog and
a he-dog together and before long they would produce a
third dog. However, place two watches side by side and they
might lie there side by side forever without producing a
third.[18]

Debate over the doctrine of animal soul had blotted a great
deal of paper by the early eighteenth century.[19] As Leonora

[16]Descartes, *Treatise of Man (Principia)* 4.1. Quoted by Thomas S. Hall, *Ideas of Life
and Matter* (Chicago: University of Chicago Press, 1969) 1:262.

[17]Jacques Roger, *Les Sciences de la vie*, 339–43.

[18]Fontenelle, *Lettres diverses de M. le Chevalier d'Her**** (1683). Cited by Shirley A.
Roe, *Matter, Life and Generation*, 1.

[19]Leonora Cohen Rosenfield, *From Beast-Machine to Man-Machine* (New York: Oxford University Press, 1941) is the best work on the subject. Her introduction to
Pardies's *Discours de la connoissance des bestes* (1672) (New York: Johnson Repr. Corp.,
1972) was also extremely helpful in my discussion of Pardies. See also George Boas,
The Happy Beast in French Thought of the Seventeenth Century (Baltimore: Johns Hopkins
Press, 1933); Jean Ehrard, *L'Idée de nature en France dans la première moitié du XVIIIe
siècle* (Paris: S.E.P.E.N., 1963) 1:673–90; Elizabeth Fontanay, "La Bête est sans
raison," *Critique* (Paris) 34 (1978): 707–29; Albert Balz, "Cartesian Doctrine and the
Animal Soul: An Incident in the Formation of the Modern Philosophical Tradition,"
Studies in the History of Ideas 3(1935):117–77.

Cohen Rosenfield has demonstrated in her study of animal automatism, *From Beast-Machine to Man-Machine,* the Cartesian view was questioned prior to 1740 both by writers who called themselves Cartesians in other respects, and by those who were open adversaries of the system. By 1734, with the publication of the first volume of Réaumur's *Histoire des insectes,* a considerable literature on the subject existed. Most writers, even those who were otherwise followers of Descartes, admitted some kind of animal soul. Rosenfield has written, for example: "Indeed, by 1747 animal automatism had lost ground with the general public. In 1737 the Cartesian Macy had openly acknowledged the defeat of the Cartesian doctrine."[20]

Of great importance in formulating an alternative to the Cartesian view of animals as soulless automata was *Discours de la connoissance des bestes* (1672) by Père Ignace-Gaston Pardies. Though operating within the Cartesian framework of ideas that equated the human soul with reason, Pardies argued that animals possess a material soul of an intermediate third substance between extended matter and the non-extended rational soul of man. Pardies rejected Descartes's view of animals as unfeeling. He thought that they were capable of both sensory knowledge and emotions. Pardies's use of the machine analogy, adding to it a soul of a "unique and indivisible substance," ruined the neat dualism that Descartes had set up. His description of animal soul reveals an ingenious modification of the Cartesian system.

The power of God will not be less admirable, since besides the many springs which compose this machine, and which causes all the parts to make the appropriate movements, He will have found the means to make a soul, which though completely material has the faculty to know and perceive objects; He will have been able to join this soul to this machine with a bond so intimate and so indissoluble that of these two parts, that is to say the body and soul, He has made a unique and indivisible substance; finally, He will have been able to fill the whole earth with an infinity of diverse sorts of animals which are on the one hand so similar to us and so

[20]Leonora Cohen Rosenfield, *From Beast-Machine to Man-Machine,* 57.

close to our nature; and on the other hand, so unlike us and so infinitely below us.[21]

Another important advocate of animal soul was Noël Régnault (1683–1763), the Jesuit author of *Entretiens physiques d'Ariste et Eudoxe,* who also argued in the second edition, which was published in 1732, that animal soul was of a third substance, intermediate between brute matter and spirit.[22] Both Pardies and Régnault continued to be popular well into the eighteenth century. Pardies had seven editions by 1724; Régnault appeared in eight editions between 1729 and 1755. Rosenfield has called both authors representative of the neoscholastic or peripatetic viewpoint, the official position of the universities and the Jesuit order.[23] The peripatetic idea of animal soul came from Aristotle, who granted a sensitive, as opposed to a rational soul, to animals. By the decade of the 1730s, the Cartesian idea of animal automatism had fallen into disrepute. It awaited Trembley's discovery to be reawakened.

Given the extent of the debate and Réaumur's natural reticence, it is not surprising that Réaumur exercised caution when discussing animal soul. In the introduction to his *Histoire des insectes* he carefully noted the problems and issues on which the debate centered.

But will we refuse all intelligence to insects and reduce them to the simple state of a machine? Here is the great question of the soul of animals, debated so often since Descartes, and about which so much has been said since the debate began. All that we owe to the disputes which it engendered is that the two opposing opinions

---

[21]"[L]a puissance de Dieu n'en sera pas moins admirable, puis qu'outre tant de ressorts qui composent cette machine, & qui en disposent tous les membres à faire les mouvemens qui leur sont propres, il aura trouvé le moyen de faire une ame, qui toute materielle qu'elle est, a la faculté de connoître, & d'appercevoir les objets; qu'il aura pû joindre cette ame avec cette machine d'un lien si intime & si indissoluble, que de ces deux parties, je veux dire du corps & de l'ame, il se fait une substance unique & indivisible; enfin, qu'il aura pû remplir toute la terre d'une infinité de diverses sortes d'animaux, qui sont d'une part si semblables à nous, & si approchans de nôtre nature; & d'une autre part si dissemblables, & si infiniment au dessous de nous." Ignace Pardies, *Discours,* 234–35.

[22]Noël Régnault, *Entretiens physiques d'Ariste et d'Eudoxe ou physique nouvelle en dialogues,* 1st ed. (Paris, 1729), 3 vols.; 2nd ed. avec des additions (Amsterdam, 1732), 4 vols.; for a critical discussion of this work see Pierre Brunet, *L'Introduction des théories de Newton en France au XVIIIᵉ siècle* (Paris: Albert Blanchard, 1931), 183–86. See also Jacques Roger, *Les Sciences de la vie,* 348.

[23]Leonora Cohen Rosenfield, *From Beast-Machine to Man-Machine,* 79–93.

both seem plausible, but it is impossible to demonstrate which one of the two is right.[24]

Réaumur argued that it was certainly within the scope of God's power to create animals capable of intelligence, but he asked would He have given the capacity of philosophizing to the sedentary oyster without putting us within reach of recognizing this intellectual activity in a creature so seemingly dull?

If someone were satisfied to argue that God could make machines capable of growing, multiplying and executing all that insects or other animals accomplish, who would dare to deny that the All-Powerful could only go this far? But if some contend that God could give to insects an intelligence equal or superior even to ours, without putting us in reach of knowing that he gave it to them; if this someone contended that an Oyster, vile that it is to our eyes, although fixed to pass on the same piece of rock, a life which seems to us very sad, could enjoy there a very agreeable life, being always occupied with the highest speculations, it cannot be denied that the supreme power can do this and more; he can create and place intelligence where he wishes.[25]

For Réaumur, the crux of the matter was whether animals exhibited behavior which suggested that they had intelligence. Réaumur reasoned that some persons might deny animal soul by pointing out that the activities of certain animals, insects, for example, are too regular. Insects do not know how to vary their actions in response to different circumstances the way humans might. However, Réaumur's research on the habits of insects revealed that some insects did indeed know how to adjust their actions when certain con-

---

[24]"Mais refuserons-nous toute intelligence aux insectes, les réduirons-nous au simple état de machine? C'est-là la grande question de l'ame des bêtes, agitée tant de fois depuis M. Descartes, & par rapport à laquelle tout a été dit dès qu'elle a commencé à être agitée. Tout ce qui a dû resulter des disputes qu'elle a fait naître, c'est que les deux sentiments opposés ne soûtiennent rien que de très-possible, mais qu'il est impossible de démontrer lequel des deux est le vrai." Réaumur, *Histoire des insectes* 1:21–22.

[25]"Mais si quelqu'un soûtenoit que Dieu a pû donner aux insectes des intelligences égales ou superieures même aux nôtres, sans nous mettre à portée de connoître qu'il les leur a données; si ce quelqu'un soûtenoit qu'une Huitre, toute vile qu'elle est à nos yeux, quoyque fixée à passer sur le même morceau de rocher une vie qui nous paroît fort triste, y peut jouir d'une vie très-agréable, étant toujours occupée des plus hautes speculations, on ne sçauroit lui nier que le pouvoir suprême ne puisse aller là & plus loin; il peut créer & placer des intelligences où il veut." Ibid.

ditions demanded it. But, Réaumur averred, "To reduce the thing to its truth, each species of insect has only, so to speak, his skill *(tour d'adresse)* through which it knows how to attract our admiration."[26] But even if insects demonstrated more varied and surprising actions or a series of actions similar to ours, "they would gain nothing in the eyes of those who are obstinately determined to refuse them souls."[27] Réaumur ended his discussion of animal soul by citing Leibniz in a rather ambiguous statement. Are not animals like ourselves, who, "viewed from the outside, act as pure machines? . . . Every human body is a machine constructed to perform a series of movements and actions, and this series is what the soul, destined to inhabit the body, wills it to perform as long as it is inhabited."[28]

If it were necessary solely to depend on Réaumur's ambiguous statements in his *Histoire des insectes* to determine his position on animal soul, it would be extremely difficult to reach any firm conclusion. However, in one instance in Réaumur's letters to Bonnet, he let down his guard and let himself reveal his true feelings on the question of soul in animals. This letter, which was discovered by Jean Torlais, was written fourteen years after the discovery of the polyp, and therefore falls outside of the period of the 1740s, which is the main focus of the present study. However, it must be cited here because it is the *only* forthright statement that has been found of Réaumur's views on the issue. On 14 March 1754, Réaumur wrote to Bonnet in a hurt tone which revealed his fundamental disagreement with Georges-Louis Buffon, the pompous *intendant* of the Jardin des Plantes, concerning the question of animal intelligence. Buffon's growing reputation after the publication of the first volumes of his *Natural History* in 1749 had at this point begun to throw the aging Réaumur's prestige in eclipse. Réaumur complained to Bonnet that "the whole misfortune of bees and other insects is that I love them and dare to admire them." Réaumur grumbled that this was enough to make Buffon and his "clique" speak of insects scornfully. "They should not, however, be embarrassed to

---

[26]Ibid., 23.
[27]Ibid.
[28]Ibid.

have them act with intelligence," he wrote candidly. "If material souls which feel their own existence can be admitted, everything can be accorded to matter. But what metaphysics!"[29]

It comes as a surprise to see that, on the issue of animal soul, Réaumur, a devoted Cartesian in questions of physics,[30] deviated from Descartes's belief that animals were mere automata, devoid of feeling, freedom, and above all of thought. Réaumur's statement of his belief that animals, and his beloved insects, in particular, could act intelligently was unusual for one who was by nature extremely reserved when it came to hurling himself into a debate that was ultimately metaphysical. However, there seems no doubt that Réaumur believed that animals exhibited genuine, if limited, rational behavior. This behavior was what had primarily intrigued him about animals, and his study of insects had focused on what he called the genius, customs, and industries of insects.[31] Réaumur uses the word genius (génie), an extremely difficult word to translate from the French, which had been employed in Aristotle to refer to the rational soul, reserved for man alone. It appears that Réaumur's use of the word was a conscious reference to the reasoned behavior of insects.

Indeed, the view that Réaumur truly believed that insects act rationally is reinforced by Bonnet's criticism of Réaumur for carrying out his belief in animal intelligence too far. Two years after Réaumur's death Bonnet wrote at Baron Melchoir Grimm's request, an analysis of Réaumur's contributions to natural history. Though acknowledging Réaumur's consummate skill as an observer of the industries of insects and his role in shaping and encouraging the contributions of his

---

[29]"[T]out le malheur des abelles et des autres insectes, est de ce que je les aime et ose les admirer. c'en est assez pour faire parler d'eux avec mepris Mr de Buffon et toute sa clique. ils ne devoient pourtant pas etre embarassés de les faire agir avec intelligence. quand on peut admettre des ames materielles qui sentent leur propre existence, on peut tout acorder a la matière. mais quelle methaphisique!" Réaumur to Bonnet, 14 March 1754, Ms Bonnet 42, Bibliothèque Publique et Universitaire de Genève. The modernized text of this comment was first published in Jean Torlais, "Un Maître et un élève. Réaumur et Charles Bonnet (d'après leur correspondance inédite)," Gazette Hebdomadaire des Sciences Médicales de Bordeaux (October, 1932), 652.

[30]The best discussion of Réaumur's Cartesian physics which were based on Jacques Rohault's Traité de physique (1671) can be found in Cyril Stanley Smith's introduction to Réaumur's Memoirs on Steel and Iron (Chicago: University of Chicago Press, 1956).

[31]Réaumur, Histoire des insectes 1:13.

disciples, Bonnet specifically criticized Réaumur for having supposed in insects "an intelligence which approximated too much our own. It happened to him, as well as to others, not to have kept enough on guard against the first impulse of admiration."[32]

It is likely that Réaumur's view of animal soul at the time of the publication of the first volume of his *Histoire des insectes* was similar to that of his friend, the renowned secretary of the Academy of Sciences, Bernard de Fontenelle, another Cartesian, who, nevertheless, challenged the doctrine of the animal machine both on the issue of generation and soul. In his work *Sur l'instinct*, Fontenelle rejected the idea of animal instinct and argued that reason and experience guide both animal and human behavior.[33] In his reviews of Réaumur's *Histoire des insectes* in the *Histoire de l'Académie Royale des Sciences*, Fontenelle interpreted Réaumur's descriptions of insects as demonstrating the balance between reason and habitual actions.[34]

Thoughout his discussions of insects, Réaumur was attracted to their intricate behavioral patterns, their ability to construct elaborate nests and cocoons, their actions taken to protect themselves from their enemies, and their foresight in providing for their young. Bonnet's rather harsh criticism of Réaumur's belief in animal intelligence, as expressed to Baron Grimm in 1759, was the product of hindsight, for, initially, it was precisely these qualities that Réaumur had observed in insect behavior which attracted Bonnet to Réaumur's work. However, what Réaumur saw as examples of the reasoned behavior of insects, Bonnet interpreted as purposive behavior which revealed the action of God's providence. For example, in his first letter to Réaumur Bonnet wrote:

I was filled with admiration at the sight of the sure and simple means which Nature employs for achieving her ends; the excellent Mechanism which rules in the different Operations of Insects, the ruses which they know how to vary when the need arises, and finally,

---

[32]Charles Bonnet, "Observations sur quelques auteurs d'histoire naturelle" in *Correspondance littéraire, philosophique et critique par Grimm, Diderot, Meister, etc.*, ed. Maurice Tourneux (Paris: Garnier Frères, 1878) 4:167.

[33]Leonora Cohen Rosenfield, *From Beast-Machine to Man-Machine*, 126.

[34]Jean Ehrard, *L'Idée de nature en France dans la première moitié du XVIIIᵉ siècle* (Paris: Ecole Pratique des Hautes Etudes, 1963), 682 and note 6.

the wisdom with which their actions seem directed, enchants me all the time. A caterpillar is now for me an animal which yields to none other in beauty . . .[35]

Bonnet's early letters to Réaumur are filled with minute descriptions of the complicated behavior of caterpillars, ant-lions, and spiders. His careful details of the behavior of the "liveried caterpillars," which Réaumur had introduced to him through the *Histoire des insectes,* were typical of these mutual interests. Réaumur admitted to Bonnet that Bonnet's description of the disciplined behavior of the liveried caterpillars, particularly their use of silk threads to assist them in finding their way back to the nest, went well beyond his own observations. This particular behavior reminded Bonnet of the threads that Ariadne used to rescue Theseus from the Cretan labyrinth. He enjoyed seeing the visible confusion of the caterpillars when the threads they needed in order to find their way back to the nest were broken:

Hardly had the first arrived at the break when, completely frightened, it retraced its steps, followed by several others who preferred to turn back rather than attempt the passage. A single one, more daring than the others, after having thoroughly explored the area surrounding this bad spot, ventured to cross it. The thread which it left in passing again established the route, but until it was entirely repaired, they always hesitated at this crossing: I have often taken pleasure in confusing one of these caterpillars by erasing all traces of the surroundings. It remains in the same place without daring to make up its mind. Finally, several threads which, naturally, I had left, put it back on its way.[36]

[35]"J'étois rempli d'admiration à la vuë des moyens également surs et simples que la Nature employe pour parvenir a ses fins, le Mechanisme excellent qui regne dans les differentes Operations des Insectes, les ruses qu'ils scavent si bien varier dans le besoin, la sagacité enfin avec laquelle leurs actions semblent dirigées, m'enchantent tous les jours; une Chenille est presentement pour moi un Animal qui ne le cede à nul autre en beauté, les Faits qu'elle m'offre au tems de sa Transformation et pendant le cours de sa vie, ont dans mon Esprit des agrémens qu'il ne sçait plus trouver dans les plaisirs ordinaires de la Jeunesse." Bonnet to Réaumur, 4 July 1738, Papers of Réaumur, Bonnet (DB), Archives de l'Académie des Sciences, Paris.

[36]"[A] peine la premiere fut—elle arrivée vers la Fracture, que toute effarouchée elle retourna sur ses pas, elle fut suivie de plusieurs autres qui aimerent mieux rebrousser que de tenter le passage; une seule plus hardie qu'elles aprés avoir bien tâté ses environs de ce mauvais pas, s'hazardât a le franchir, le fil qu'elle laissa en passant retablit le Chémin.—mais jusques à ce quil fut entierement reparé, elles hesitoient toûjours a ce passage. J'ai souvent pris plaisir à embarrasser une de ces Chenilles en effaçant toutes les traces des environs, elle restoit à la même place sans oser se determiner, enfin quelques fils que javois sans doute laissé, la remettoient dans la Voye." Ibid.

Bonnet's second letter to Réaumur, written 3 November 1738, again revealed his paramount interest in purposive or goal-oriented insect behavior. Bonnet observed a caterpillar which lived on hazel trees, which Réaumur had described in Volume One of his *Histoire des insectes*.[37] According to Réaumur, this caterpillar was among those that produced extremely large cocoons in proportion to the small size of the pupa. As in his observations of the liveried caterpillars, Bonnet experimented with the behavior of the pupa by interfering with its habitual activities, attempting to prove the animal's ability to respond to the unforeseen by the exercise of will. He found that if he touched the cocoon with the tip of his finger, the pupa would draw backwards and downwards into the unused portion of the cocoon, then suddenly it would quickly reascend to its former position in the cocoon.

It was easy to perceive the difference in the movements which she made to descend from those which she made to ascend. The former were heavy, forced movements which made me feel that the pupa was constrained to descend by a force which she could not resist. The latter, on the contrary, were voluntary movements; it was easily [seen] by the promptitude and the lightness with which she ascended that she did not force them at all, that she had a goal, and that she tended to take herself to the place from which I had chased her.[38]

The early letters which Bonnet and Réaumur exchanged, beginning in 1738 up to the discovery of the polyp, do not reveal any substantial disagreement or tension. It was only *after* Trembley's discovery that metaphysical issues, particularly the problem of animal soul, surface. However, this is not to say that there were no fundamental differences in approach. From the beginning Bonnet looked for teleological explanations, while Réaumur tried to avoid them whenever

---

[37]Réaumur, *Histoire des insectes* 1:516.

[38]"Il etoit aisé d'apercevoir la difference des mouvemens qu'elle se donnoit pour descendre de ceux qu'elle se donnoit pour monter. Les premiers etoient des mouvemens lourds, forcés, qui faisoient sentir que la Crisalide etoit contrainte à descendre par une force à laquélle elle ne pouvoit resister: Les seconds au contraire, etoient des mouvemens volontaires, on [voioit?] aisément par la promptitude et la legereté avec laquélle elle montoit; qu'elle n'y etoit point forcée, qu'elle avoit un but, qu'elle tendoit à s'aller placer à l'endroit d'où je l'avois chassée." Bonnet to Réaumur, 3 November 1738, Papers of Réaumur, Bonnet (DB), Archives de l'Académie des Sciences, Paris.

possible. Réaumur's preoccupation with the rational soul of insects, though no less real, was kept firmly in check by his focus on observation and experiment.

Just as Réaumur and Bonnet did not conceive of animal soul in precisely the same way, though both rejected the Cartesian view of the mechanical generation of animals, neither naturalist accepted epigenesis. Since the view of animals as little machines was an unquestioned premise on which the study of insects was predicated, it was necessary to espouse a theory that would address and circumvent the difficulties which the Cartesian explanations of generation and animal soul presented.

The idea of a preexistence of germs filled this need. According to the proponents of the preexistence theory, what appeared as generation was only the mechanical playing out of the program which had been fixed by God at the Creation of the world. Since God placed extremely small germs in the progenitors of each species, the only real generation was consequently the act of Creation. What appeared as generation was merely the mechanical development of infinitely small germs, encased one within the other. When circumstances, such as the union of male and female, triggered an expansion in size, the germs developed into embryos to which the female gave birth. When the original supply of germs placed in the female (or male) of each individual pair at Creation was used up, life on earth would cease.[39]

Since the equal reproductive role of male and female was not fully understood prior to the development of cell theory in the early nineteenth century, the problem that intrigued naturalists, was whether the germs were carried by the male or the female. The ovists gave the female the dominant reproductive role. For them the egg was the vehicle for the preformed individual. In mating, the male merely stimulated the growth or expansion in size of the miniature individual. In contrast, the animalculists believed that the sperm carried the miniature; the function of the egg was nutritive. The preexistent germs were also thought to be the vehicles for the soul of each animal.

---

[39]Superb discussions of the preexistence theory can be found in Jacques Roger, *Les Sciences de la vie*, 325–453.

The idea of the preexistence of germs was a product of the late seventeenth century and, as Peter Bowler has pointed out, cannot be seen as a continuation of the old preformation concepts of Giuseppe degli Aromatari, Fortunio Liceti, and Emilio Parisano.[40] These early proponents of preformation did not believe in the idea of *emboîtement,* or the encasement of germs, which is the hallmark of the later preexistence theory. In the old preformation theory, procreation resulted from the development of the fetus produced from the soul of one parent; preformation was limited to the development of parts taken from within the parent and did not extend backwards to include previous generations. However, it had in common with the later preexistence theory the view that, prior to conception, a preformed miniature existed in the body of one parent.

Though the doctrine of preexistence that spread after 1670 shared with the earlier view the idea that generation was really only a development of preformed parts, the new theory added that the

germ contained in the seminal fluid or in the seed is not produced by the parent: it was created by God himself at the beginning of the world, and conserved from that time until the moment of its "development." The adult who seems to have engendered it or brought it into the world served in reality only as its shelter or nurse.[41]

The idea of a preformed miniature in the unfertilized egg did not depend on new observations made possible by the scientific use of the microscope. Rather, it filled the philosophical need to perceive the world as the product of the rational design of God. This is why the preexistence theory fitted so well into the movement of ideas called insecto-theology.

Neither Anton van Leeuwenhoek nor Marcello Malpighi, the two greatest microscopists of the late seventeenth century,

---

[40]Peter Bowler, "Preformation and Pre-Existence in the Seventeenth Century: A Brief Analysis," *Journal of the History of Biology* 4 (1971):221–44; Jacques Marx, "La Préformation du germe dans la philosophie biologique au XVIII siècle," *Tijdschrift voor de Studie von de Verlichting* 1 (1973): 397–428; and Jacques Roger, *Les Sciences de la vie,* all note this discontinuity between the early preformation theory and the idea of preexistence of germs of the late seventeenth century.

[41]Jacques Roger, *Les Sciences de la vie,* 326.

claimed to have observed the embryonic preformed individual within the sperm or egg. Leeuwenhoek, though he leaned toward the idea of male preformation after he discovered animalcules in the male seminal fluid in 1677, did not state that he had actually seen the preformed miniature in the spermatozoon. It was left to the less careful microscopist, Nicholaas Hartsoeker, to develop the idea of the male "homunculus."

Moreover, though later preexistence theorists used Marcello Malpighi's observations on the chick embryo as justification for the doctrine, this was, in fact, a misinterpretation of Malpighi's observations which were made on the fertilized egg prior to incubation. Though an ovist, Malpighi had never claimed to have observed the "fetus" of the chick in the unfertilized egg. Malpighi recognized that there was nothing in the unfertilized egg that might be called a miniature embryo.[42]

An important contribution to the acceptance of the idea of preexistence was the increased support for the view that the egg or ovum was the vehicle for procreation, and a decline in the animalculist argument that the spermatozoa were responsible for generation. The discovery of the ovum in mammals was claimed by Régnier de Graaf in 1672. Though, in fact, he had not actually discovered the ovum, but only the follicles of the ovary from which the eggs emerge, his claims gave impetus to the developing ovist position.[43] Eggs of the viviparous mammal were thereafter seen as analogous to those of an oviparous animal. In addition, the seeds of plants were thought to have the analogous preformed potential.

By the early eighteenth century Swammerdam was generally cited as the authority for the view that the embryo preexisted in the egg prior to fertilization. Indeed, there were hints in Swammerdam's work on insects that provided grist for the proponents of the theory. He claimed that the "fetuses" in the eggs of insects were alive and endowed with feeling prior to fertilization.[44] The biblical reference that

[42]Peter Bowler, "Preformation and Pre-Existence," 226.

[43]Jacques Roger, Les Sciences de la vie, 262.

[44]As the authority for this idea, in one of his letters Pierre Lyonet cites the 1685 French translation of Swammerdam, 60; 57, lines 2 & 3; 58, lines 16 & 17; 81, line 15; 94, lines 19 & 20; 138, lines 16 & 17. I could not obtain this edition, so was not able to check these references.

Swammerdam used to argue against chance generation—that Levi paid tithes to Melchisedech several generations before Levi was conceived—clearly could be interpreted in a way that would lend credence to the idea of preexistence.[45] Moreover, the specific reference in the *Miraculum Naturae* of 1672 that all humans develop from eggs contained in the womb of Eve clearly indicates Swammerdam's attraction to the theory.[46]

Swammerdam's greatest contribution to entomology, his dissections of the stages in the life cycle of the butterfly, proved that the complete metamorphosis claimed by William Harvey was only an appearance. He carefully observed and described in meticulous detail the development of the parts of the body of a later stage within the insect's earlier stage. The stages in the life cycle, in appearance so completely different, were the outward manifestation of an internal orderly progressive development from fertilized egg to caterpillar to pupa to butterfly. Swammerdam's emphasis on the fixed regularity of insect development gave support to the mechanical image of nature. It is not difficult to see how, in the imagination of the time, it was but a small step to extrapolate the idea of the development of germs from his evidence of the regular development of the butterfly beneath the apparent miraculous transformation from worm-like creature to the ethereal winged insect. Was this not proof of, or at least analogous to, the development from germs of the animal machine, enveloped within the womb of its progenitor from the time of creation?

Swammerdam's work was considered of great importance by naturalists of the early eighteenth century. Boerhaave purchased Swammerdam's papers and published the *Biblia naturae* in 1737–38. Trembley, Bonnet, and Réaumur all held him in high regard. Bonnet paid Trembley the highest com-

---

[45]The payment of tithes to Melchisedech is mentioned in Genesis 14:18–20.

[46]Bowler is not convinced that Swammerdam was fully committed to the preexistence concept. "It is evident that Swammerdam was fully aware of the pre-existence concept, but it would be difficult to reconcile the belief that he became committed to the complete *emboîtement* theory with his own insistence on the epigenetic development of insects as a means of eliminating the operations of chance from nature. This would imply that he neglected his own work in order to adopt a belief which he obviously found interesting, but which he never expressed as more than a brief speculation." Peter Bowler, "Preformation and Pre-Existence," 238.

pliment when after discovery of the polyp he wrote that Trembley walked in the footsteps of the great Swammerdam, "for whom the least Things were marvels."[47]

If the observations of Swammerdam, Malpighi, and de Graaf did not actually confirm and justify a belief in preexistence, they did contribute to the philosophical construction of the theory, which seems to have received its most complete statement in the work of Nicholas Malebranche. Recognizing the limitations of the Cartesian theory with regard to generation, Malebranche specifically set forth the full theory of *emboîtement*, or the encasement of germs, in the first volume of his *Recherche de la vérité*, published in 1674. Malebranche reasoned that a single seed of an apple tree might contain an infinite number of seeds to supply an infinite number of centuries. Since it was possible to demonstrate mathematically the divisibility of matter to infinity, "that is sufficient to have us believe that there could be smaller and smaller animals to infinity, though our imagination is shocked at this thought."[48] Malebranche stressed the inability of the senses as well as the imagination to conceive of such infinitely small objects of creation.

Roger's assessment of the importance of Malebranche's influence stresses that his thought was a reflection of the main currents of the thought of his time.[49] The theory of *emboîtement* was a means of adjusting the mechanical vision of the world to the belief in God's role as both Creator and Architect. Moreover, it was a rationally consistent explanation of generation. The organized body, a term which is constantly used from the late seventeenth century, was the living body. Within the organized body were an indeterminate number of little organized bodies that were born through time. Unorganized matter was incapable of organization. The gulf between non-living brute matter and the organized being was unbridgeable.

Descartes had supposed that the inert matter of the male and female semen could organize itself. But, Malebranche

---

[47]Bonnet to Trembley, 24 March 1741, George Trembley Archives, Toronto, Ontario.

[48]*Recherche de la vérité* (Paris, 1938), 73. Cited by Jacques Roger, *Les Sciences de la vie*, 336.

[49]Ibid., 337.

objected: "One can well believe that the general laws for the communication of motions suffice for the development and the growth of the parts of organized bodies; but one cannot be convinced that they can ever form a machine which is so composed."[50] The infinite complexity of the organized body could never have been a product of chance.

The preexistence theory of Malebranche, however, was not the only one that was proposed to save biological mechanism. Also emphatically against the Cartesian theory of generation and probably influenced by Perrault, Wilhelm Gottfried Leibniz proposed a theory of preexistence which differed radically from that of Malebranche. The Leibnizian solution involved a different conception of matter.[51] Rather than passive and inert, as in the Malebranchian view, following so closely that of Descartes, Leibniz believed that all matter, and not merely that making up the organized body, is dynamic and organized. Life is everywhere, even in the smallest conceivable particles of matter. In Descartes's theory an animal was merely a functional unity, which had life as long as the parts worked together acccording to mechanical laws. Malfunction of a part could cause death and return to inert, unorganized matter, just as a careless slip in the cutting of one wheel could cause a watch to fail to work. As Malebranche put the matter: only a watch which the watchmaker has finished putting together works.[52] Regardless of the perfection of each piece, without the rational integration of each carefully calibrated part into the whole, the machine would not do the work for which it was designed.

In contrast, for Leibniz, there could be no rigid separation between the organic and the inorganic realms. Matter was never passive and inert, but dynamic and alive. The problem was how to impose a unity on this mass of living matter to result in the animal machine. To guarantee the necessary unity of the organism, a soul was indispensable. Leibniz declared, "I count as corporal substances only the machines of

---

[50]Quoted from Malebranche, *Entretiens sur la métaphysique et sur la religion*, by Paul Schrecker, "Malebranche et le préformisme biologique," *Revue Internationale de Philosophie* 1 (1938): 81.

[51]Jacques Roger, "Leibniz et les sciences de la vie," *Akten des internationalen Leibniz Kongresses* (Wiesbaden, 1969) 2:209–19.

[52]Paul Schrecker, "Malebranche et le préformisme biologique," 89.

nature which have souls or something analogous, otherwise there will be no true unity at all."[53] Leibniz believed that germs or souls were created when God created the world. The material bodies to which they were attached, however, were subject to continual transformations. Generation and death were only appearances, "developments and envelopments of which nature shows us visibly some samples as is its custom, to aid us in guessing what it hides."[54] Generation did not depend on the passive vehicle of the egg, which enclosed the germ in Malebranche's view of preexistence, to insure continuity of the germ from Creation to the present. Rather, in his mature thought, Leibniz was convinced that the spermatozoon carried the germ, and so he supported Leeuwenhoek in the face of attacks by the ovists, Louis Bourguet and Antonio Vallisnieri.

Leibniz believed, contrary to those who refused the spermatozoist argument because of the waste it entailed, that a sperm which did not develop into an animal did not die, but continued to live as a simple organism. Generation was a kind of transformation of the preexistent germ attached to the spermatozoon. There was no real death in nature, since the soul, as well as the animal to which it was connected, continued to exist, although the visible forms might change. "Thus, not only the souls, but also the animals, are ungenerated and imperishable: they are only developed, enveloped, reclothed, stripped, transformed; souls never completely leave their bodies."[55]

For Leibniz, preexistence was related to his idea of continuity, which was embodied in the belief in the great chain of being or scala naturae. He believed that all of creation, both the organized and apprently unorganized, was connected by imperceptible degrees. There were no jumps or lacunae. The idea of continuity supported his view of matter as dynamic and everywhere alive. Animal, vegetable, and mineral realms could not be rigidly separated, as in the Cartesian system, where there is an unbridgeable gap between the organized, hence living body, and brute matter. Leibniz's scale of nature

---

[53]Leibniz, letter to Jaquelot, 22 March 1703. Cited by Jacques Roger, "Leibniz et les sciences de la vie," 212.

[54]Letter to Lady Masham, May 1704. Quoted by Jacques Roger, ibid., 213.

[55]Leibniz, Principes de la nature et de la grâce. Cited by Jacques Roger, Les Sciences de la vie, 370.

seemed to imply a kind of transformism, since species closely
related to each other on the scale might *seem* to have developed from older ones. Leibniz emphatically denied this. The
chain, though it might seem to be progressively realized
through time, was completely static. It was part of the pre-
established harmony, laid down at Creation. However, not
all animals and plants were visible or "developed" at the same
time.[56]

The preexistence theory as it developed in the late seventeenth century is of importance in understanding the
thought of Bonnet, Trembley, and Réaumur. All three have
been called "preformists," though this designation is more
complicated than historians such as Cole have realized in
their discussions of the scientific contributions of Trembley
and Bonnet. In addition to distinguishing the preexistence
of germs which involves encasement from simple preformation, it is important to recognize the distinction which
Jacques Roger has drawn between Leibnizian preexistence
implying an underlying dynamic view of matter, and a passive
emboîtement of the miniature within the egg. Réaumur appears
to have followed Malebranche in his view of encased germs
within the egg. Bonnet, however, may have been influenced
by the Leibnizian view, though support for this argument
remains indirect, bound up in the fabric of the science which
he practiced, rather than seen in any explicit declaration.
That is to be expected. After all, at this time Bonnet was an
observer. Yet observation is selective. It is shaped by factors
which are never fully articulated. It is to these factors, which
are at least in part cultural, that we will turn in the following
chapter.

---

[56]Arthur D. Lovejoy's analysis of Leibniz and the chain of being in *The Great Chain
of Being* (Cambridge, Mass.: Harvard University Press, 1936) has been extremely
helpful in my understanding of these concepts.

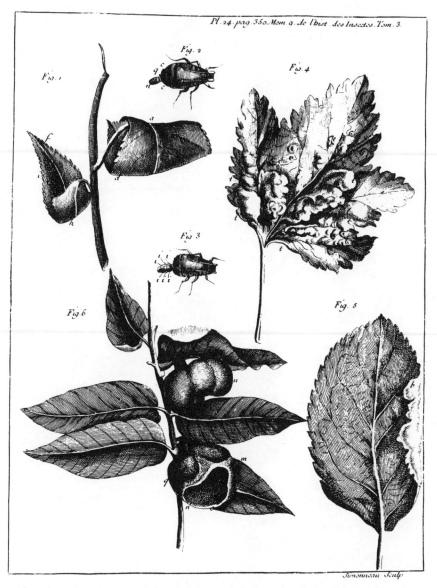

Plate from Réamur's *Mémoires pour servir à l'histoire des insectes,* vol. III shows an aphid in the process of giving birth (fig. 2–3). Also shown are the various ways that different species of aphids use leaves to construct shelters for themselves.
(Permission to reproduce from Réamur's *Mémoires,* courtesy of the Rare Book Collection of The Cleveland Health Sciences Library, Historical Division.

Chapter 3

Geneva: The Cultural Matrix

I. THE RELIGIOUS CONTROVERSY AND THE INTRODUCTION OF CARTESIAN SCIENCE

Before looking in greater detail at the education that Bonnet and Trembley received at the Academy, it is necessary to consider the historical background within which their science was to develop in order to understand their intellectual development and to place their letters in perspective. Because of its geographical situation in the heart of Western Europe, Geneva was a crossroad for travelers and a center for the exchange of ideas. And because it was Protestant in Catholic Southern Europe, Geneva had ties with Protestants in distant countries, particularly Holland and the north German states. Easy access to the Rhine River made trips to Holland particularly convenient. Europeans traveling to Italy on the Grand Tour usually stopped in Geneva and the city, through a sizable indigenous Italian population, served as a conduit for Italian culture to pass across the Alps into Northern Europe. A close relationship had also traditionally been kept with England, a tie that prompted one eighteenth-century contemporary to quip, "Geneva is a town which speaks and writes French, but thinks and reads English."[1] Facility with English was certainly true of Trembley, who was an able translator. The city had been one of the havens for French Protestants fleeing religious persecution throughout the seventeenth century and especially after the revocation of the Edict of Nantes. The Protestant Bonnet family left France after the St. Bartholomew Massacre in 1572. The Trembley family was

[1]Béat de Fischer, "Swiss in Great Britain in the Eighteenth Century" in *The Age of Enlightenment*, W.H. Barber, ed. (Edinburgh: Oliver and Boyd for University Court of the University of St. Andrews, 1967), 352.

also of French origin. The family left France twenty years
earlier than the Bonnet family, during the reign of Henry
II. It is not certain whether they left because of their Protestant beliefs. The religious climate at that time, though uncertain, was not openly hostile.

Proud of its history as an independent republic, which
lasted until the end of the eighteenth century,[2] though connected geographically and linguistically to France, Geneva
never submitted to the intellectual hegemony of the French
Enlightenment. In contrast to the atheism and deism which
prevailed in eighteenth-century France, organized religion,
strongly supported by the educational and political institutions, was able to keep a balance between reason and revelation.

Genevan theologians such as Jacob Vernet (1698–1789)
incorporated belief in natural law and a critical and historical
approach to the Bible into Calvinism. However, the Genevans' faith in man's reason which, according to the historian
Ernst Cassirer, was the "unifying and central point of this
century,"[3] was not a simple importation of ideas of the French
philosophes. It was this assumption that so infuriated Genevans
when D'Alembert's controversial article in the *Encyclopédie*
appeared in 1757. Probably prompted by Voltaire, who had
recently taken up residence near Geneva, d'Alembert de-
picted the city as the near epitome of enlightened government. Geneva's only fault, he declared, was its proscription
of theater. D'Alembert cast eighteenth-century Calvinist theology as Socinian or Unitarian and questioned whether the
clergy still felt revelation was "necessary." D'Alembert wrote:

In short, many of the ministers of Geneva have no other religion
than a perfect Socinianism; they reject everything called "mystery"
and imagine that the first principle of a true religion is not to
propose any belief that conflicts with reason. . . . Respect for Jesus

---

[2]Though Geneva formed military alignments with other Swiss states and with
France for protection, Geneva did not join the Confederation of Swiss States until
after the Congress of Vienna in 1815. This was the result of bitter experience under
Napoleon.

[3]Ernst Cassirer, *The Philosophy of the Enlightenment* (Boston: Beacon Press, 1965),
5.

Christ and for the Scriptures is perhaps all that distinguishes the Christianity of Geneva from pure deism.[4]

The rationalism of the clergy, which d'Alembert misinterpreted, was the result of the integration into Genevan Calvinist theology of a pattern of ideas that served to break down the scholastic theology of the seventeenth century. The challenge came not from the *Philosophes* of the eighteenth century, but from the gradual penetration of a new liberal theology from within Calvinism. The defeat of the old scholastic Calvinism with its heavily Aristotelian emphasis came in the last decades of the seventeenth century, and coincided with the gradual introduction of the "new science" of Galileo, Gassendi, Copernicus, and Descartes into Genevan thought. Thus, from the beginning, the "new science" was connected with a more liberal theology, and the close relationship between science and theology continued throughout the eighteenth century.[5]

Though science and religion were not autonomous anywhere in Europe in the late seventeenth and early eighteenth centuries, their relationship in Geneva remained particularly close because of the power of the clergy—power which was embodied in the institutions set up by John Calvin. Foremost among these institutions was the Academy of Calvin, founded in 1559. The Academy was not only the training ground for future Protestant ministers, but also prepared Genevan youth for positions in the administration of the Republic. Calvin had drawn up not only the ecclesiastical code, but also the civil one, and he stabilized the governmental structure, which consisted of four syndics, nominated by the Small Council and elected by the Council of Two Hundred. In addition, the Small Council proposed legislation, which was voted on by the Council of Two Hundred.

Though not a theocracy, since the Small Council, which did not consist only, or even primarily, of clergy, had ultimate

---

[4]Jean le Rond d'Alembert, "Geneva" in *Encyclopedia* (Indianapolis: Library of Liberal Arts, 1965), Nelly Hoyt and Thomas Cassirer, eds., 138.

[5]Michel Heyd's study, *Between Orthodoxy and the Enlightenment: Jean-Robert Chouet and the Introduction of Cartesian Science in the Academy of Geneva* (Boston: International Archives of the History of Ideas 97, 1982) emphasizes that the acceptance of Cartesianism and the new science was a process of gradual accommodation, not confrontation.

authority on matters of church doctrine, strong clerical influence in the Republic was assured through the control of the Academy by the clergy. Since the Academy produced Geneva's public officials, it was inevitable that the clergy would have a large role in shaping the ideas and attitudes of those who served the Republic. The Academy was ruled by a group of men drawn from the churches of the city and called the Venerable Company of Pastors. The Venerable Company controlled the nomination of professors in all but a handful of cases where laymen were allowed to hold university appointments. These were proposed by the Small Council. The Venerable Company controlled the curriculum as well, an important aspect of its power when the introduction of science was later debated.

Though Geneva was a republic in name, in fact it was an oligarchy, controlled by a handful of patrician families, known as the citizens—the class at the top of a rigidly hierarchical class structure. Both Trembley and Bonnet were citizens and later in life looked upon their service to the Republic as members of the Council of Two Hundred both as a duty and a privilege. The citizens were native-born sons of citizen families. They were the only class allowed to become magistrates.

The bourgeois class also enjoyed a large measure of participation in the governance of the city. Like the citizens, they could vote and be elected to the two governing councils. A bourgeois was either the son of a citizen, but born outside of Geneva, or was the son of a bourgeois. In some cases the son of a bourgeois could become a citizen. In addition, a few foreigners were granted the rights of the bourgeois. The two lower classes, the residents and the natives, had no rights to participate in the governance of the city. Residents were foreigners residing in the city; natives were sons of residents who enjoyed some additional privileges that residents could not receive. According to D'Alembert's article in the *Encyclopédie,* written in the middle of the eighteenth century, the population of the city was about 24,000. Of this number, about 1,500 adult males had the privilege of voting.[6]

---

[6]D'Alembert, "Geneva," 130.

After the death of John Calvin, the leadership of the Academy passed to his disciple, Theodore Beza (1519–1605). Beza's influence is reflected in the increasingly scholastic emphasis of the teaching of the Academy, which became one of the leading training centers for the Protestant ministers of Europe. Beza was an Aristotelian and deeply influenced by the Italian Aristotelians of the School of Padua, particularly Pietro Pomponazzi.[7] The Aristotelianism of orthodox Geneva was not a unique European phenomenon. Other centers of Protestantism, such as Leiden and Amsterdam, also reflected this Aristotelian revival. Why we find in the sixteenth century a movement away from Calvin's strong emphasis on personal religious experience and the Bible has not been fully explained by historians. Calvinism may have become increasingly scholastic as a response to the need to explain and justify Protestant doctrines attacked by Catholic theologians.

For Beza the central doctrine of the Church became that of Election. Beza declared that election preceded the Fall, and salvation was granted only to a select few. Beza presented this doctrine in a scholastic manner. It was part of a logically coherent system that had the effect of removing Christ and salvation through grace from the central place it had held in Calvin's theology.[8] The installation of Aristotle in theology brought with it Aristotelian science and at first delayed the acceptance of the new science of Copernicus, Galileo, and Descartes.

The first challenge to Beza's rigid orthodoxy came from the Dutch theologian Jacob Arminius (d. 1609). Arminius preached against the harshly exclusive doctrine of predestination, which denied salvation to all but the Elect. Arguing in the anti-scholastic dialectic of Peter Ramus, Arminius preached that salvation was given for all men, though not all men chose to accept it.[9] Arminius stressed the ethical di-

---

[7]In addition to Theodore Beza, scholasticism in Reformed theology is exemplified by two Italians, Girlamo Zanchi and Peter Martyr Vermigli. Aristotelianism developed in Renaissance Italy simultaneously with humanism, drawing its inspiration from the medieval scholastic traditions of Paris and Oxford. See Brian Armstrong, *Calvinism and the Amyraut Heresy* (Madison: University of Wisconsin Press, 1969), 127–28.

[8]Ibid., 38–42.

[9]Indeed, Beza had denied a teaching post to Peter Ramus in the Academy of Calvin. Ibid., 38.

mension of human experience. He asked rhetorically, if only
a few were elected, and if election did not depend on moral
behavior, what incentive did it provide to live virtuously?
Were not the majority of men damned despite their efforts
according to Beza's doctrine of election? Arminius argued
that grace "depends on the will of man, in regard that by
virtue of its native liberty, it may receive or reject this grace,
use it or not use it, render it effectual or vain."[10] In effect,
Arminius claimed that through good works humanity could
earn salvation. This was heresy because it placed a limitation
on God's power and freedom. To combat the growing threat
of division within Calvinism that Arminianism posed, the
representatives of the Reformed Churches met at the Synod
of Dort in 1618. There they reaffirmed the orthodox position.

However, a more liberal Calvinist tradition continued to
be evident. Though the Academy of Calvin in Geneva remained a bastion of the orthodox, academies in France, allowed to flourish because the Edict of Nantes had granted
religious toleration, offered an alternative to the harsh doctrines of Geneva. The new liberal Calvinist theology, which
ultimately penetrated Geneva as well, was forged by John
Cameron and Moïse Amyraut at the Protestant Academy of
Saumur.[11] This was known as the covenant doctrine and represented a compromise between the emphasis on free will in
the Arminian doctrine and the blind determinism of the orthodox position.

The central teaching of John Cameron, who was called
"Beza's scourge," was that of a covenant between God and
man. The covenant was based on love and restored the importance of Christ as an instrument in bringing grace. Cameron's emphasis on the covenant removed predestination
from its old position as a central doctrine. It gave greater
scope to the use of man's reason. *Illuminatio* or conversion
required the use of reason to accept or reject the offer of
grace. "God does not convert the sinner by blind acceptance;

---

[10]Perry Miller, *The New England Mind: The Seventeenth Century* (Boston: Beacon
Press, 1961), 368.

[11]The best discussion of the Saumur theology is Brian Armstrong, *Calvinism and
the Amyraut Heresy.*

rather He illumines and demonstrates His truths to the intellect with such force that the will automatically and necessarily shows adherence."[12] Thus, it was through reason that the sinner was led to faith. Paradoxically, though the divine truths were ultimate mysteries, the path to achieving grace was through the exercise of human reason.

Moïse Amyraut, Cameron's successor, continued to build upon the covenant theology of his teacher. Amyraut declared that there were two covenants, one secret and hidden, but also unconditional and absolute, the other revealed, which carried with it the condition of faith. Since all men do not fulfill the condition of faith, Amyraut called the doctrine "hypothetical universal grace," which derived from the second, or conditional covenant. All men have the potential to be saved, but all men do not fulfill the condition of faith. Amyraut's doctrine of "hypothetical universal grace" was called "Arminian" by the orthodox Calvinists. Amyraut was charged with heresy in 1637 and 1644. This doctrine came to be known as the "Amyraut heresy," and the clash over it almost destroyed the Academy in Geneva.

The Amyraut heresy, while undermining the authority of Aristotle, encouraged the empirical use of reason. Amyraut denied that theology could be a logically consistent system, deduced from first principles, as Beza had proposed, for to accept that view was to presume to have knowledge of God's purpose. He argued that man must start from what is observable and build up a concept of God empirically.[13] If man applies his powers of observation in a rational manner, he will discover that the universe is the product of rational design. The external world is governed by a system of natural laws put in place by God. These natural laws are an example of God's providence. Predestination is simply that part of providence which concerns man.

Most interesting is Amyraut's use of the watch analogy to demonstrate the providential ordering of the world. Give an atheist a watch, he argued, and ask him if it was made by itself, if the springs and wheels could have been placed to-

---

[12]Walter Rex, *Essays on Pierre Bayle and Religious Controversy* (The Hague: Nijhoff, 1965), 95.
[13]Brian Armstrong, 179.

gether by chance? The answer is obvious. In the same way
that a "cause endowed with intelligence" has intervened in
the craftsmanship of a watch, the construction of the world
demonstrates God's creative power "to move this whole ce-
lestial machine so constantly for so many centuries."[14] In
Amyraut the mechanical world view takes its place incon-
gruously alongside pre-Copernican motion of the sun to
demonstrate the rationality of God's creation.

The precise relationship between the Saumur doctrines
and the importance of the design argument has not been
studied, but it raises several interesting questions. Why did
Swammerdam in 1663 interrupt his medical studies at Leiden
to spend a year at Saumur, where he is said to have demonstrated the valves of the lymphatic vessels? What is the
connection between the Saumur doctrine, the design argument and the official acceptance of Cartesianism there in
1664? Is there a connection between the belief of the Cambridge neo-Platonists that "reason is the candle of the Lord,"
and the concept of *Illuminatio* common to both Cameron and
Amyraut?[15]

Though these larger questions cannot be answered, the
history of how these distinctive teachings of the Saumur
school with its liberal theology insinuated themselves into the
orthodox stronghold of Geneva can be seen in the tensions
which it produced within the Academy of Calvin between
1639 and 1708. This process had political as well as theolog-

---

[14]"Car tout le monde sçait que ce sont les hommes que le font, & qu'estant
impossible que le hasard ait joint ensemble tant de parties avec tant d'art, il faut
necessairement que l'operation de quelque cause doüée d'intelligence y soit inter-
venuë. De là venés à leur demander s'ils croyent qu'il y ait moins d'art en la con-
stitution de [sic] monde qu'en celle d'une montre; Si le mouvement d'eguille qui
marque la distinction des heures, est plus reglé que celuy du soleil qui les fait; si
les roües dont le mouvement depend sont plus artificielles que les spheres celestes:
si le ressort qui les fait tourner est mieux & plus reglement tendu, pour les faire
mouvoir vingt-quatre ou trente heures seulement, que la puissance qui meut si
constamment depuis tant de siecles? toute cette machine celeste?" *Sermon sur le verset
I. du Pseaume XIV* (Saumur, 1645). Quoted by Brian Armstrong, 274. Here the
mechanical world view is espoused prior to the acceptance of the Copernican revolution.

[15]See Ernst Cassirer, *The Platonic Renaissance in England*, tr. James Pettegrove
(Austin: University of Texas Press, 1953) and Rosalie Colie, *Light and Enlightenment,
A Study of the Cambridge Platonists and the Dutch Arminians* (Cambridge, England:
Cambridge University Press, 1957).

ical overtones and was complicated by the introduction of Cartesianism.

The religious controversy reflected the political division between the bourgeoisie who allied themselves with the new liberal theology, and the citizens, who supported the orthodox Calvinist position. During the struggle, Geneva was subjected to the external pressure of the Evangelical Cantons of Bern, Zurich, Basel and Schaffhausen which supported the orthodox side. Also involved was the theological issue of a critical approach to Scripture by the liberals, an approach opposed by the orthodox conservatives who insisted on the literal truth of the Bible.

The first indication of the approaching conflict with the Academy came in 1639, when an individual suspected of teaching the new liberal Saumur doctrines was appointed by the Small Council to a professorship.[16] His appointment was supported by the Small Council against the Venerable Company of Pastors. Thereafter, the Venerable Company forced every candidate for the ministry to sign a statement in which he promised to support the canons of the Synod of Dort and specifically rejected the new doctrine of universal grace. This rule was called the *sic sentio,* and different forms of it were implemented in 1647, 1649, 1658, and 1659.

The conflict between the orthodox and liberal theology simmered until June 1669, when two professors of theology in the Academy, Louis Tronchin, representing the Saumur doctrines, and François Turrettini, adhering to the official dogma, openly clashed over the issue. Ironically, both had received a part of their training at the Saumur Academy. However, while Turrettini used his knowledge of the Saumur doctrines the better to oppose them, Tronchin became an ardent supporter of Amyraut.[17]

François Turrettini is given credit for holding liberal Calvinism at bay for thirty years until his son Jean-Alphonse Turrettini succeeded in legitimizing the teaching of the new doctrines in 1708. However, though officially the teaching of

---

[16]Charles Borgeaud, *Histoire de l'Université de Genève: L'Académie de Calvin (1559–1798),* (Genève: Georg, 1900), 353–56.

[17]After his return to Geneva, Louis Tronchin corresponded with Amyraut. See Charles Borgeaud, 357; Heyd, 152.

the Saumur doctrines was forbidden, Walter Rex's study of
the correspondence of Pierre Bayle, a student at the Academy
at the time of the 1669 controversy, shows that hypothetical
universal grace continued to be taught in Tronchin's home
where he invited students for private courses.[18]

The effect of the controversy was extremely destructive.
Traditionally the teaching of theology had been the Academy's most important discipline and had attracted students
from all parts of Switzerland. But because of the dispute, the
conservative cantons did not encourage their candidates for
the ministry to attend the Academy. For example, an investigation by the Small Council found that while before 1669
there were approximately fifty-five students, by 1673 this
number had been reduced to seventeen in theology. Moreover, Tronchin and his colleague in theology, Philippe Mes-
trezat, used any pretext not to lecture. When they did lecture,
they excised the portions of their lectures which dealt with
the forbidden subjects, leaving their lectures disorganized
and incoherent.[19]

Though in one respect the year 1669 was a dark one for
the Academy, in another respect it was a turning point. In
the fall of that year Jean-Robert Chouet (1642–1731), a
young professor of philosophy at the Academy of Saumur,
arrived to fill the chair left vacant by the death of his former
professor of philosophy at Geneva, Gaspard Wyss. This appointment was of great importance for the fortunes of the
Academy and for the development of science in Geneva.
Chouet, of an important Genevan publishing family, was the
nephew of Louis Tronchin. As their correspondence shows,
Tronchin was mentor and friend as well.

Chouet introduced Cartesianism into the Academy of Calvin. With the new rational philosophy came the complete
acceptance of the new science and the final rejection of the
geocentric universe which had been supported by the weight
of Aristotelian physics and Biblical authority. Thus, with the
acceptance of Chouet's Cartesianism, the tight deductive logic

---

[18]See Walter Rex, *Essays on Pierre Bayle*, 138ff. Rex points out that Turrettini was
not a reactionary and his theology was not a simple continuation of the old scholastic
tradition of Beza. See also Elizabeth Labrousse, *Pierre Bayle* (La Haye: Nijhoff, 1963),
103ff.

[19]Heyd, 153; Borgeaud, 362.

of Beza's scholasticism lost its appeal without direct confrontation with the Venerable Company over doctrinal issues.

It was Tronchin who had succeeded in having Chouet appointed to the Academy without the usual competition required of candidates for chairs. Chouet's appointment gave an advantage to supporters of the liberal theological position, though the Venerable Company may have thought that their position was secure, since Chouet had signed a statement denying belief in the Saumur doctrines and promising that, if the occasion arose, he would teach only the orthodox position. Chouet was careful in his teaching to avoid the forbidden subjects, so that the issue never was discussed in public.

Chouet's own introduction to Cartesianism had come in 1662 through David Derondon (or De Rondon) with whom he studied at the Protestant academy at Nîmes.[20] In 1664 Chouet took Derondon's ideas to Saumur where he successfully competed for the chair in philosophy against the Aristotelian candidate. Chouet remained at Saumur until he was called to Geneva, blossoming into a skillful teacher with a large student following. It appears that at first he was reluctant to return to Geneva, writing in 1667 to his uncle that he was in the perfect location to learn both from Catholics and Protestants, and he was close to Paris "which is absolutely the source of men of letters, and by whom I am informed of an infinity of things regarding the sciences, of which I would probably be ignorant if I were elsewhere."[21]

When Chouet came to Geneva in 1669, his teaching met with the same brilliant success he had enjoyed at Saumur and it appears that many of his students from the Saumur Academy followed him. Not only did Chouet bring with him students who were probably tainted with the Saumur philosophy, but after the revocation of the Edict of Nantes in 1685, Geneva was temporarily inundated with French Protestants fleeing persecution for their religious beliefs. It is likely that they brought with them elements of the Saumur heresy which had by then spread to other Protestant academies. However,

---

[20]Expelled from France for a polemic against the mass, David Derondon gave private lessons in theology at Geneva in 1663, probably arranged by Louis Tronchin. He died there in 1664. See Eugène Arnaud's *Notice sur David de Rondon, professeur de philosophie à Die, Orange, Nîmes et Genève* (Nîmes, 1782).

[21]Eugène de Budé, *Vie de Jean-Robert Chouet*, 51.

as Michael Heyd points out, Chouet never openly clashed
with the Venerable Company over theological issues and may,
in fact, have viewed the controversy over the Amyraut heresy
with a degree of skepticism as to its importance.[22]

Heyd stresses that the introduction of Cartesianism by
Chouet was not as radical a break with the Aristotelianism
of orthodox Calvinists as previous authors, such as Charles
Borgeaud, have argued. Elements of the new science had
already begun to influence the teaching of philosophy, undermining the scholastic edifice well before Chouet returned
to Geneva. Chouet's Cartesian philosophy had the advantage
of appearing as a "coherent and systematic" alternative to
the more eclectic and contradictory thought of his predecessor.[23] Gradually, through Chouet's leadership, the exclusive power of the Venerable Company over education was
weakened. In 1708 Chouet and Jean-Alphonse Turrettini
were instrumental in restructuring the Academy, giving the
lay professors positions in the newly-created academic senate.

Turrettini, the student and protégé of Chouet, led the
reform of Calvinism which laid the foundation for the rationalism of eighteenth-century Geneva. His teachings were
based on a belief in the rational design of the universe by
God and a critical approach to Scripture. For his role in the
reform of theology he became known as the "new Calvin."
As Heyd has perceptively stated, the new theology, through
its emphasis on natural law, encouraged the study of nature
and was similar to the natural theology common in England.

Natural theology, in fact, assumed a distinct and major role in
Turrettini's teaching, serving as a firm and essential basis for the
exposition and proof of revealed Christian religion. The central
concept in his natural theology was that of natural law, both in its
physical and in its moral sense. . . . Turrettini, thus, provided an
ideological justification for the study of nature, a justification which
was current in England already in the seventeenth century, but
which Chouet, at this time, was careful not to stress.[24]

---

[22]Heyd, 55–56.

[23]Ibid., 134.

[24]Ibid., 200–201. Richard S. Westfall, *Never at Rest: A Biography of Isaac Newton*
(Cambridge, England: Cambridge University Press, 1980), 532, mentions that Turrettini called on Newton in 1693 bringing him news of Nicolas Fatio de Duillier's
illness. Fatio is presumably the first Genevan convert to Newton's views. Heyd does
not discuss possible Newtonian influence on J.-A. Turrettini.

Chouet's scientific legacy to the Academy of Calvin can be seen in his successful effort to have science officially recognized by the Academy by the creation of a chair of mathematics. At first this chair was only honorary. It was filled without significant influence on the Academy by Etienne Jallabert, who was related by marriage to Chouet. However, in 1724 the chair of mathematics was given formal recognition by the joint appointment of Gabriel Cramer and Jean-Louis Calandrini, both of whom, it has been pointed out, were teachers of Trembley and Bonnet. Through the political influence of Chouet, the Small Council appointed them without the concurrence of the Venerable Company. Both had been students of Chouet, picked by him to carry on the teaching of science which he had begun. By an ingenious arrangement, they were allowed to alternate several years of teaching with those of travel, thereby maintaining close contacts with the rest of the European scientific establishment. Because of their alternating teaching schedule, they became known as the Castor and Pollux of Genevan science. In addition to Bonnet and Trembley, among the students of Cramer and Calandrini were Théodore Tronchin, Jean Jallabert, Georges Louis Le Sage, Jean Antoine Butini, Louis Necker, and Louis Bertrand, all of whom became scientists of reputation.[25]

When Cramer and Calandrini were appointed, Jean-Alphonse Turrettini was the rector of the Academy. He enthusiastically supported the creation of the chair, and encouraged students to engage in the new studies. "Do you wish to sharpen your mind, increase its capacity, render it truly exact and precise in research? Go to the mathematicians, the geometers, and give yourselves to their discipline. There is the true logic, the true school of reasoning."[26]

The importance of the history of the struggle which brought about the liberalization of theology in Geneva lies in the fact that its ultimate triumph was accompanied by the relatively late introduction of the new science. With the gradual substitution of the mechanical universe for the qualitative

---

[25]Cleopatra Montandon, *The Development of Science in Geneva in the XVIIIth and XIXth Centuries: The Case of a Scientific Community*, 119–20.

[26]Pierre Speziali, "Gabriel Cramer et ses correspondants" in *Conférence faite au Palais de la Découverte* (Université de Paris, 6 December 1958), 2–3.

worldview of Aristotle, the integrity of the scholastic theology was undermined without direct confrontation. Once Aristotelian science had been called into question, a new theology was mandated. However, the accommodation of the theology taught at the Academy of Calvin to the new ideas was only slowly accomplished. While in France the new science could flourish in the Academy of Sciences independently of the established university system, in Geneva the Academy *was* the university. The institution adjusted slowly to the new ideas of the scientific revolution.

The Cartesian science introduced from France by Chouet was not resisted because Chouet chose to emphasize, not the revolutionary implications of the new philosophy, but its more traditional aspects, such as how well it could be accommodated to the "sober philosophy" of a practicing Christian. In his wish to avoid becoming embroiled in controversy, and possibly also because from the beginning not all the Cartesian doctrines accommodated themselves well to Protestant beliefs, Chouet carefully avoided the more polemical aspects of Cartesian metaphysics. This possibly made it easier for the Genevans to be influenced by new intellectual currents. Particularly important in the shaping of the science of the second generation of scientists was the influence of Newton and Leibniz. In contrast, by the late seventeenth century Cartesianism had taken a firm hold on French minds. The Protestants, who might have kept the intellectual environment in France more fluid, had been driven out and their academies closed. Cartesianism became the official philosophy of the Jesuits and had important influence on education through Jesuit schools. While Réaumur's education by the Jesuits was strongly Cartesian, the education that Bonnet and Trembley received in Geneva reflected the crossfertilization of ideas that the religious controversy had produced.

II. EDUCATION UNDER CRAMER AND CALANDRINI

Jean-Robert Chouet's determination to make sure that Genevan science did not develop in isolation from the main

European intellectual currents probably contributed to the eclectic nature of Genevan scientific thought. As students of Chouet, Cramer and Calandrini were firmly grounded in the Cartesian science of their predecessor. But the gifted exemplars of the new generation greatly expanded the ideological base on which this science rested by combining influences from two seemingly incompatible sources: the natural philosophy of Isaac Newton and the mathematics of his celebrated Continental rival, Gottfried Wilhelm Leibniz.

Because Geneva shared a common Protestant background with England, the influence of British natural philosophy was felt much earlier in Geneva than in France. Genevans were acquainted with the works of Locke and Newton before they penetrated the mountain passes "on the brilliant and light wings of Voltaire."[27] The attraction of Newtonian science was in part theological. Newton's system provided an alternative to the harsh mechanical world of Descartes. Newton's suggestion that gravity did not have a mechanical cause, but was evidence of God's providential role in nature seems to have appealed to Cramer and Calandrini. They were devout Christians who placed their scientific teaching in a religious context. According to Bonnet, among students their authority was all the more respected since they were laymen on a faculty composed almost entirely of ministers: "what they said in favor of the Revelation did not fail to strike the students and did not contribute in small measure to forewarn them against the dangerous sophisms of incredulity."[28]

Initially, it appears that what attracted Cramer and Calandrini to Leibniz was not his metaphysics, but his mathematics. In May 1727 Cramer was given his first opportunity to travel, visiting Basel, Leiden, London and Paris. He lost no time in establishing his first professional scientific connection with a group of Leibnizian mathematicians in Basel, known as the Bernoulli circle. Cramer spent six months there

[27]"Locke et Newton furent connus à Genève avant qu'ils eussent, comme dit Cousin, 'franchi le détroit sur les ailes brillantes et légères de l'imagination de Voltaire.' Et, de bonne heure, leur influence se fit sentir dans l'Ecole. Les idées de Locke pénétrèrent les spéculations de métaphysique, celles de Newton les constructions des physiciens." Charles Borgeaud, *Histoire de l'Université de Genève*, 563.

[28]Charles Bonnet, *Mém. aut.*, 44. Ms 2856 Bibliothèque Publique et Universitaire de Genève, "Essai sur la religion naturelle," by Calandrini illuminates his extreme piety.

studying principally with Johann (I) Bernoulli (1667–1748). This was a singular honor since the elder Bernoulli accepted only the most promising students. In Basel Cramer met other members of the Bernoulli circle: Leonard Euler, and Daniel and Nicholas Bernoulli, establishing lifelong friendships with each. Johann Bernoulli and his brother Jacob (I) (1654–1705) can be regarded as the major avenue by which Leibnizian physics and mathematics were introduced into Switzerland.

Evidence of the high esteem in which Cramer was held by Johann Bernoulli was his selection as editor of the two elder Bernoullis' works.[29] According to Jacob Vernet, the Genevan theologian who wrote Cramer's "Eloge historique" in the *Nouvelle Bibliothèque Germanique,* "Long since initiated in the mysteries of the infinitely small and in the great questions which then divided the best Mathematicians of Europe, Mr. Cramer acquired in Basel knowledge which would have put him in a good position to enter into this dispute if he had wished."[30] According to Vernet, Bernoulli had urged Cramer to write a work defending *vis viva,* but Cramer had refused saying that he did not want to begin his career with a polemical work.[31] Vernet reported that when Cramer visited London he did, however, propose several experiments designed to throw light on the famous question.

From Basel Cramer went to Leiden and during his extended stay there formed in important scientific contact with William Jacob 'sGravesande, whom he visited from July to December 1729. This established the association of the Genevans with the Leiden Newtonians, among whom 'sGravesande and Herman Boerhaave figured prominently.[32]

[29]For more information on the connection of the Bernoullis with Leibniz, see *Dictionary of Scientific Biography* (New York: Scribners, 1970) 2: 52. Cramer edited Johannis I Bernoulli, *Opera omnia* (Geneva, 1742), 4 vols.; Jacobi I Bernoulli, *Opera* (1744), 2 vols., and Leibnitzii & Bernoulli, *Commercium epistolicum philosophicum et mathematicum* (1745), 4 vols. Cramer's letter, 9 September 1737, to Jallabert, Ms S.H. 242, fols. 20–21, Bibliothèque Publique et Universitaire de Genève, reveals his enormous respect "pour tout ce qui s'apelle Bernoulli." For additional biographical material on Cramer, see also Jean Torlais, *Réaumur,* 175–79.

[30]Jacob Vernet, "Eloge historique de Gabriel Cramer," *Nouvelle Bibliothèque Germanique* 10 (1752): 365.

[31]Ibid., 366.

[32]Pierre Brunet, *Les Physiciens hollandais et la méthode expérimentale en France au XVIIIᵉ siècle* (Paris: Librairie Scientifique Albert Blanchard, 1926). See also Robert Schofield, "An Evolutionary Taxonomy of Eighteenth-Century Newtonianisms," *Studies in Eighteenth-Century Culture* 7 (1978): 179–80.

The Dutch grafted Newton on to a well established Baconian empiricism, and this orientation influenced Cramer. The experimental tradition of Newton of the *Opticks,* can be clearly seen in a letter from Cramer to Jallabert in 1738 when his student was visiting 'sGravesande. Cramer wrote that he was convinced that in optics refraction could be explained as an effect of attraction. The more he thought about it, Cramer wrote, "the more I am persuaded that attraction is a fact which must have a mechanical cause. I am ignorant of it, but I recognize its existence."[33]

'sGravesande inspired Cramer's only published book, *Introduction à l'analyse des lignes courbes algébriques* (Geneva, 1750). According to the theologian Jacob Vernet, Cramer took Newton's work as a starting point, but "it was Mr. 'sGravesande who inspired this work most, and who informed of these discoveries, assured him that he had found the key to the complete Theory."[34]

The strong scientific connection between Geneva and Holland reinforced their religious and political ties. Both were small Calvinist republics. However, while experimental science flourished at Leiden, the faculty of theology was more conservative then its counterpart in Geneva. Well into the eighteenth century, Leiden theologians supported the rigid interpretation of predestination laid down at the Synod of Dort in 1618. Because 'sGravesande's *Introductio ad philosophiam, metaphysicam et logicam continens* (1736) supported the idea of the ethical freedom of the individual, it was vigorously attacked.[35] Thus, the Arminian heresy appears to have resurfaced in Holland in a scientific work.

In 1729 both Cramer and Calandrini were encouraged by 'sGravesande to contribute to his *Journal Littéraire.* Published between 1712 and 1722 in ten volumes, it had suspended publication. Anxious to have foreign contributions to the newly reorganized publication, 'sGravesande asked not only

---

[33]Cramer to Jallabert, [1738], S.H. 242, fol. 39r, Bibliothèque Publique et Universitaire de Genève.

[34]Jacob Vernet, "Eloge historique de Gabriel Cramer," 382. Curiously, Pierre Speziali in "Gabriel Cramer et ses correspondants," 20, observed that Cramer may never have mastered the infinitesimal calculus since it is not used in the *Introduction* or in his correspondence.

[35]Mouza Raskolnikoff, "De L'Education," 88.

for literary news, but also for short articles. Just what they published there remains to be investigated, though it is suspected that this journal will be important as a source for a more complete understanding of the science of Cramer and Calandrini. Place of publication was moved from The Hague to Leiden in 1732 where it appeared under a new name, *Journal Historique de la République des Lettres,* until 1733 when publication ceased completely.[36]

Cramer and Calandrini were also active in founding the periodical *La Bibliothèque Italique* in 1728. They collaborated in this undertaking with the philosopher of Neuchâtel, Louis Bourguet (1678–1742), a committed Leibnizian who had close ties with Geneva. This was one of the first journals in Switzerland to publish the work of local scholars. Previously, they had contributed to periodicals like the *Journal Littéraire,* published in Holland.[37] Printed in Geneva in a series of eighteen volumes until 1734, it introduced French-speaking readers to the importance of Italian culture. Long articles featured the contributions of the Italians to science, particularly the experimental work in the natural history of insects of Marcello Malpighi, Giacinto Cestoni and Antonio Vallisnieri. In addition to drawing attention to recently published books by Italian authors, literary news from cities like Rome, Bologna, Florence, Venice, as well as Geneva was included. Longer articles featured excerpts from the *Mémoires de l'Académie des Sciences de l'Institut de Bologne* and letters of a polemical nature, such as the arguments of the Newtonians against the vortices of Descartes and their defense. Thus, the journal hoped to stimulate the interest of other European scholars in the lively intellectual life which Bourguet had found in Italy between 1710 and 1715.

---

[36]For further information on 'sGravesande and this journal see: *Oeuvres philosophiques et mathématiques de Mr. G.J. 'sGravesande,* ed. Jean Nicholas Sébastien Allamand (Rey, 1774). Calandrini's "Dissertation sur la force des corps" is reprinted there, 269–73. Margaret C. Jacob in *The Radical Enlightenment: Pantheists, Freemasons and Republicans* (London: George Allen and Unwin, 1981) sees a connection between the literary society at The Hague from which the *Journal littéraire* emerged and Freemasonry. She points out that despite his connections with the more radical pantheistic philosophy of his associates, 'sGravesande remained faithful to the more conservative Newtonian providentialism, 185–87.

[37]See André Sayous, *Le Dix-huitième Siècle à l'étranger* (Paris: Amyot, 1861; repr. Geneva: Slatkine, 1970), 140.

Obviously, Calandrini, whose family was of Italian origin, must have found this collaboration extremely satisfying. Though less well known, possibly because of his reticence to publish, he had a reputation in Geneva equal to that of Cramer with whom he maintained a close friendship. During his life, he was chiefly known as a commentator on Newton's *Principia*.[38] He corresponded with Alexis-Claude Clairaut on the theoretical questions raised by Newton's theory of the moon's apsis. Bonnet reported in his *Mémoires autobiographiques* that Calandrini published a small *Traité des sections coniques* which was included in the commentaries on Newton's *Principia* edited by Thomas Le Seur and François Jacquier.[39]

At the Academy Calandrini generally taught algebra and astronomy, while Cramer taught geometry and mechanics. When Calandrini was elected to public office in 1750, Cramer took over the chair of philosophy. It should be emphasized that until 1737 when Jean Jallabert (1712–68), himself a student of Cramer and Calandrini, was appointed honorary professor of experimental physics, instruction by Cramer and Calandrini was the *only* means by which a student could obtain formal instruction in science in Geneva.[40]

It is likely that Trembley and Bonnet received similar, though not identical educations at the Academy. While Calandrini was the dominant influence on Trembley, Cramer played the major role in Bonnet's education. However, given the peculiar terms of Cramer and Calandrini's joint appointment in which they alternated years of teaching with travel, Trembley could not have escaped Cramer's influence as well. One significant difference between their educations may have been that of timing. Trembley completed his dissertation in

---

[38]Charles Bonnet, *Mém. aut.*, 45. The full title of this three-volume work is *Philosophiae naturalis principia mathematica* (Geneva: Barillo & Filii, 1739–42). It was known as the "Jesuit's Edition" of the *Principia* and, in addition to Calandrini's commentary, contained important papers on the theory of the tides by Daniel Bernoulli and Leonard Euler. See Craig B. Waff, "Universal Gravitation and the Motion of the Moon's Apogee: The Establishment and Reception of Newton's Inverse-Square Law, 1687–1749," (Ph.D. Diss., Johns Hopkins, 1975), 111–123.

[39]Charles Bonnet, *Mém. aut.*, 45.

[40]Cramer's letters to Jallabert, 1737–40, Ms S.H. 242, fols. 20–90 Bibliothèque Publique et Universitaire de Genève, are an important source for a description of the influences on Genevan experimental science. For additional background on Jallabert, see Isaac Benguigui, ed., *Théories électriques du XVIIIᵉ siècle* (Genève: Georg, 1984).

1730. Bonnet, ten years younger, did not begin his work in the Auditoire de Philosophie until 1736.

Trembley's studies, unfortunately, have not been described in detail. He had planned to study theology but an attack of the smallpox left him in delicate health. Did he actually begin these studies before 1733 when he left for Holland? Was he a student of Jean-Alphonse Turrettini whose liberalization of the severe Calvinist dogmas had coincided with the establishing of the chair of mathematics? This seems likely. Trembley's interest in theology was lifelong and ultimately led in his old age to several publications in which he reaffirmed his belief in natural and revealed religion.[41]

A notebook in the private archives of George Trembley in Abraham Trembley's hand with quotations from a variety of authors on religious subjects provides insight into Trembley's theological preoccupations and, in particular, his predilection for British authors concerned with the design argument and the action of God's providence. This influence of British natural theology is not surprising, in view of the strong Genevan connection with Protestant thought in England already established through the teachings of J.–A. Turrettini. The notebook has 162 numbered pages, some of which appear to have been removed. Although the notebook is not dated, it contains on its flyleaf a careful record of thermometric observations which begins with an entry dated 12 November 1738 and ends 27 January 1740. These dates indicate that the notebook was in Trembley's possession during the years of his research on the polyp on Count Bentinck's estate near The Hague. Since it is certain that many of the notes were made after 1736 (the date of publication of one of the works consistently cited by Trembley: Joseph Butler's, *The Analogy of Religion*), it is likely that at least some of the notes were made after his departure from Geneva and cannot be considered part of his formal education there.

However, it is highly likely that the philosophy of Samuel Clarke, one of the authors most consistently and extensively cited by Trembley in his notebook, was indeed part of the

---

[41]Abraham Trembley, *Instructions d'un père à ses enfans, sur la nature et sur la religion* (Geneva: Chapuis, 1775), 2 vols.; and *Instructions d'un père à ses enfans, sur la religion naturelle et révélée* (Geneva: Chirol, 1779), 3 vols.

general or theological curriculum at the Academy in Geneva. Bonnet, though he does not include Clarke among the authors he read as part of the formal curriculum he followed under Cramer and Calandrini, was familiar with Clarke's philosophy. In addition to mention of the Leibniz-Clarke correspondence in his letter of 21 October 1768 (quoted above, p. 21), Bonnet in an earlier letter to Trembley included Clarke in his pantheon of great men of the century: "Europe no longer produces names celebrated in metaphysics and philosophy: gone are Locke, Leibnetz [sic], Clarke and s'Gravesande [sic]. It still had a Volf [sic] and death has carried him off."[42]

Trembley's notebook indicates that he carefully studied Clarke's sermons which are extensively quoted under such headings as "God (wisdom of)," "Eternity," "Predestination," "Imperfections." Most likely the work which Trembley studied was *A Demonstration of the Being and Attributes of God: more particularly in Answer to Mr. Hobbs, Spinoza and their followers, wherein the Notion of Liberty is Stated, and the Possibility and the Certainty of it Proved, in Opposition to Necessity and Fate: Being the Substance of Eight Sermons Preach'd at the Cathedral of St. Paul in the year 1704 at the lecture founded by the Honorable Robert Boyle, Esquire.*[43] This work is popularly known as his Boyle Lectures. In these sermons Clarke used Newtonian science to discredit the materialists, Hobbes and Spinoza, and the mechanical philosophy of Descartes.[44] He denied that matter and motion alone were sufficient to account for all the phenomena of nature. The active intervention of God through providence was necessary to explain certain facts for which no adequate scientific explanation could be given. For example, Clarke, following Newton, argued that gravity had no mechanical explanation. Gravity, therefore, was evidence of God's continuing, active presence in nature. It was this characterization of the universe as requiring God's continuing active intervention that led to Leibniz's charge in the Leibniz-

---

[42]Bonnet to Trembley, 28 July 1761, George Trembley Archives, Toronto, Ontario. The Volf referred to is probably Christian Wolff (1679–1754), interpreter of Leibniz.

[43]I am not sure what edition Trembley used.

[44]See Gerald R. Cragg, *Reason and Authority in the Eighteenth Century* (London: Cambridge University Press, 1964), 33ff.

Clarke correspondence that the Newtonians' God had not created a perfect universe, but rather that it was necessary for Him from time to time to adjust its mechanisms.

Clarke's philosophy in the Boyle Lectures was not entirely consistent, since it contained implicitly two arguments for the being and attributes of God. The first, the *a priori* proof, based on quasi-mathematical deductive reasoning, was a logical demonstration of God's existence, using a chain of propositions. The second proof was *a posteriori*, proceeding from what was given in experience (the phenomena of nature) to its cause, the Designer or First Cause. This was a teleological argument which took observation as its starting point to build the inductive argument that nature gave evidence of intelligent planning.[45]

The design argument had, in fact, been implied by Newton himself in the *Opticks*, Query 28 (1706 Latin edition). Clarke's Newtonian philosophy, particularly his objections to the Cartesian image of a completely mechanical universe and his argument for the continuing, active presence of God in his Creation must have been what appealed to Trembley who was brought up in the devout atmosphere of Calvinist Geneva. Though the specific connection with the Newtonian view of providence, the Amyraut heresy, and the "new Calvinism" preached by Turrettini is not yet clear, it is easy to see how Newton's provision for the continued active presence of God within His Creation was part of the appeal of his natural philosophy.

Happily, in contrast to the sketchy details of Trembley's education, Bonnet left a detailed account of his education at the Academy in his *Mémoires autobiographiques,* an autobiography written as a series of letters to his closest friends, Albrecht von Haller, Abraham Trembley and Horace-Bénédict de Saussure. Bonnet's passion for insects, first awakened by his reading of Pluche's *Spectacle de la nature* had caused him to seek out Cramer and Calandrini. Cramer, Bonnet reported, had also read Pluche and thought highly of the work.

---

[45]See discussion by John Dahm, "Science and Religion in Eighteenth-Century England: The Early Boyle Lectures and the Bridgewater Treatises," Ph.D. Dissertation, Case Western Reserve University, 1969, 49. See also James P. Ferguson, *The Philosophy of Dr. Samuel Clarke and its Critics* (New York: Vantage Press, 1974).

Cramer then recommended that Bonnet read Régnault's *Entretiens physiques d'Ariste et Eudoxe*. This, as has been pointed out, was a Cartesian work, but one which deviated from a strict Cartesianism by including a discussion of animal soul. Bonnet read this work more than once and he admitted that what appealed to him was its simple style which enabled him effortlessly to assimilate scientific facts. "It seemed to me that this book put me in possession of all the parts of science *[la physique]* with the least cost possible and that I had only to read to become erudite *[savant]*."[46] After reading Fontenelle's *Entretiens sur la pluralité des mondes* several times which left a "profound impression," he turned to William Derham's *Physico-Theology or a Demonstration of the Being and Attributes of God from his Works of Creation,* a work of natural theology whose notes were crammed with examples drawn from Newton's *Opticks.*[47] Though probably Bonnet was ignorant of the fact, much of Pluche's *Spectacle* was derived from Derham's *Physico-Theology.* Both works placed great emphasis on the argument from design. They added a Newtonian orientation to Bonnet's studies which supplemented the Cartesian background of Fontenelle and Régnault. Now he was ready for the *Logique de Port-Royal,* known as *L'Art de penser,* by Antoine Arnauld. Bonnet read this more than once and though he liked the chapters on the practice of logic, he was repelled by the excessively scholastic aspects of the work. He found the logic of Calandrini in a small manuscript volume which his professor lent him more to his liking, and he attended a private course given by Calandrini on physics and speculative philosophy.[48]

Although it is not mentioned in the *Mémoires autobiographiques,* Bonnet must have taken a course in logic under Cramer

---

[46]Charles Bonnet, *Mém. aut.,* 46.

[47]Derham's *Physico-Theology* first appeared in 1713. Twelve English editions appeared before 1754. French editions appeared in 1726, 1730 and 1732.

[48]Maurice Trembley collected Bonnet's complete library parts of which subsequently have been sold. In the George Trembley Archives there is a letter from the Genevan bookseller Baumgarten and Cie that indicates possession of two unpublished volumes by Calandrini: *Logices. Cours de Logique.* L'Académie de Genève recueilli par Charles Bonnet, Pars I et III, 2 vols. Ms; *Cours de métaphysique.* L'Académie de Genève recueilli par Charles Bonnet. Ms de 277 feuilles in 8 en latin. These were presumably bought by Maurice Trembley, though I did not find them in the George Trembley archives.

as well, since a 348-page manuscript volume, *Cours de logique* by Gabriel Cramer, prepared by Charles Bonnet, exists in the George Trembley archives.[49] In this work Cramer claimed that good logic demonstrated that philosophy and theology could not be incompatible. No proposition that was false in philosophy could be true in theology. With good logic one could recognize that "the Truth is one and that the most sublime dogmas of the faith might well be above reason, but that they cannot be contrary to reason."[50] Cramer was critical of Descartes's delusion in believing that he could envision a completely mechanical world and dispense with God's providential governance of the world:

> Mr Descartes seems to have wished to use the same line of reasoning to explain all the phenomena of the universe; he took the things explainable by matter and motion, and began to examine what followed from this supposition. Soon his fertile imagination derived from it the elements, and built from it the vortices, imagining stars of suns in the center of these vortices . . . It is a pity that such a beautiful idea is nothing but a fiction, easy to destroy: or rather, it is good that a System which leads minds which are badly instructed to believe that they can dispense with the Government of Providence in the Administration of the world, can be regarded as a fiction by anyone who does not let himself be taken in by the ardor of a beautiful imagination, but who wishes to weigh a little the consequences.[51]

Newton's science may have been viewed by Cramer as the necessary antidote for the Cartesian "excessively fertile imagination." The Newtonian concept of God's active role through providence was a welcome tonic for the wholly mechanical

[49]This may be the same *Cours de logique,* mistakenly thought to be by Rousseau, sold in London in 1945. See *Dictionary of Scientific Biography* 3:462, note 1.

[50]Gabriel Cramer, *Cours de logique,* L'Académie de Genève recueilli par Charles Bonnet, George Trembley Archives, Toronto, Ontario, 267.

[51]"Mr Descartes semble avoir voulu suivre la même route pour expliquer tous les Phenomènes de cet Univers, il prit des choses connues de la matière & du mouvement, et se mit à examiner ce qui résulteroit de cette supposition. Bientot son imagination feconde en forma les Elements, en bâtit des Tourbillons, lui fit voir des Etoiles ou des Soleils au centre de ces Tourbillons . . . Il est facheux qu'une si belle idée ne soit qu'un Roman aisé à détruire: ou plûtot, il est bon qu'un Système qui meneroit des Esprits mal faits à croire qu'on peut se passer du Gouvernement d'une Providence dans l'Administration du Monde, ne puisse être regardé comme un Roman par quiconque ne se laisse pas entrainer à la fougue d'une belle imagination mais qui veut peser un peu ses consequences." Ibid., 134.

universe. However, Cramer did not recommend the *Principia* or the *Opticks* to his eager student. He may have sensed that Bonnet, though he could appreciate Fontenelle's "geometrical spirit" would have been repelled by any but the most superficial treatment of physical science. Voltaire's *Eléments de la philosophie de Newton* (1738) was therefore selected by Cramer, which he supplemented with a small manuscript commentary which he prepared himself.

It seems clear that if the common elements in Trembley's and Bonnet's education are considered, a prominent place must be given to their interest in works which can be considered quasi-Newtonian. The Newtonian character of Clarke, Pluche, and Derham, derives from their use of the design argument. The purpose of observation of nature is to unveil God's design, the art and purposes for which the world was created. Revelation is written in nature's objects: it is for the observer to decode them. For Clarke, providence could be seen in the action of gravity, which he regarded as a subordinate instrument of God. Derham carried this idea further by insisting that gravity was imprinted on matter at Creation.[52] Of greater significance for Derham was the imprinting by God of instinct in animals. The study of instinct in animals revealed the direct participation of God in his Creation. Since animals, according to Derham, could not act rationally, yet exhibited regular patterns of behavior, animal instinct could be understood only as examples of providential intervention.

And lastly, what less than Rational and Wise could endow animals with various Instincts, equivalent in their special way to Reason it self? In so much that some from thence have absolutely concluded that these Creatures had some glimmerings of Reason. But it is manifestly Instinct, not Reason they Act by, because we find no varying but that every Species doth naturally pursue at all times the same way, without any Tutorage or Learning: whereas *Reason*, without instruction, would often vary, and do that by many Methods, which Instinct doth by one alone.[53]

In his characterization of animal instinct as providential, Der-

---

[52]See John Dahm's discussion, *Science and Religion in Eighteenth-Century England*, 61–63. See also William Derham, *Physico-Theology*, 31, note 1.

[53]William Derham, *Physico-theology*, 214.

ham was drawing out the implications of Newton's questions in the *Opticks* which were couched in terms of the design argument. Newton wrote:

How came the Bodies of Animals to be contrived with so much Art, and for what ends were their several parts? Was the Eye contrived without Skill in Opticks, and the Ear without Knowledge of Sounds? How do the Motions of Body follow from the Will, and whence is the Instinct in Animals?[54]

Derham included the migration of birds among the original instincts such as food gathering imprinted in animals by the Creator.[55] The study of the response of insects to light made by both Bonnet and Trembley was a way to study instinctive behavior. For example Trembley compared the ability of polyps to move to a previously lighted spot to the ability of birds to migrate.[56] Bonnet remarked on the effects of a lighted candle on the behavior of his caterpillars.[57]

Bonnet reported that while he was still a student in philosophy he read Malebranche's *Recherche de la vérité* several times. It was Malebranche's description of the encasement of germs, one within the other, which attracted him, as well as Malebranche's love and knowledge of insects:

Malebranche knew insects, he loved them and knew how to admire them; he had penetrated the beauties of nature and the adorable perfections of its Author; he was also the great apostle of the preexistence of germs; and I was struck with this vigor of genius which allowed him to imagine that all the apple trees which have existed, which exist and will exist, were originally contained with their trunks, their roots, their branches, their twigs, their leaves, their flowers and their fruits in the first seed of this tree. I was enchanted with this world in miniature, no larger than a musket ball which contained in this prodigious abridgment all the beings which peopled the great world we inhabit. . . . It is to be regretted that the

---

[54]Isaac Newton, *Opticks* (New York: Dover, 1953), repr. 4th ed., 1730, Query 28, 369–70.

[55]William Derham, *Physico-Theology*, 318–19.

[56]Maurice Trembley, *Correspondance inédite entre Réaumur et Abraham Trembley* (Geneva: Georg, 1943), 48–49.

[57]Bonnet to Réaumur, 4 July 1738, Papers of Réaumur, Bonnet (DB), Archives de l'Académie des Sciences, Paris.

nearly mystical theology of this great man so often spoils his philosophy.[58]

As an antidote to Malebranche's "mystical philosophy," Bonnet read at Cramer's suggestion three solid empirical works in natural history: Jan Swammerdam's *Bible de la nature* and Marcello Malpighi's *L'Anatomie des plantes* and his *Dissertation sur le ver à soie*. It was through Jallabert, visiting 'sGravesande in December of 1738, that Cramer obtained Swammerdam's book, which Cramer and Bonnet read together.[59]

It should be noted that it was only after the study of Swammerdam's book that Bonnet chanced upon Réaumur's *Histoire des insectes* in 1739, thus reinforcing the developing empirical orientation of his studies. Bonnet began the study of aphids, or plant lice, in May of 1740.[60] There is no indication that Réaumur proposed by letter that Bonnet apply himself to this problem. Réaumur did not communicate by letter to Bonnet between January and June, a fact which suggests that Bonnet undertook this research on his own initiative or that of his professors. He had read Réaumur's description of the aphid in Volume Three of his *Histoire des insectes*. Réaumur stated that both Leeuwenhoek and Giacinto Cestoni had regarded the aphid, or plant louse, as hermaphroditic and he admitted that he himself had never seen a coupling. This he regarded as strange, because, unlike bees which may mate in their hives, the aphid is always within view on the stalks or

---

[58]"Malebranche connaissait les insectes, il les aimait et savait les faire admirer; il était pénétré des beautés de la nature et des perfections adorables de son Auteur; il était encore le grand apôtre de la préexistence des germes; et j'étais frappé de cette vigueur de génie qui lui faisait concevoir que tous les pommiers, qui ont existé, qui existe et qui existeront, étaient originairement contenus avec leurs troncs, leurs racines, leurs branches, leurs rameaux, leurs feuilles, leurs fleurs et leurs fruits dans le premier pépin de cet arbre. J'étais enchanté de ce monde en miniature, pas plus grand qu'une balle de mousquet, qui contenait dans ce prodigieux raccourci tous les êtres qui peuplent le grand monde que nous habitons. . . . Nous avons bien à regretter que la théologie presque mystique de ce grande homme gâte si souvent sa philosophie." Charles Bonnet, *Mém. aut.*, 93.

[59]Cramer to Jallabert, 24 December 1738, S.H. 242 Bibliothèque Publique et Universitaire de Genève.

[60]See See H. Ehrard, "Die Entdeckung der Parthenogenesis durch Charles Bonnet" *Gesnerus* 3 (1946): 15–27; Jean Rostand, *La Parthenogenèse animale* (Paris: Presses Universitaires de France, 1950).

leaves of plants.[61] According to Réaumur, Swammerdam had
stated that insects became fertile and mated only after their
final metamorphosis, but Réaumur suspected that aphids
might be an exception to this rule. Through dissection, well
developed fetuses could be observed in the bellies of aphids
prior to the appearance of their winged stage. To test this
idea, he selected an aphid of a species that he knew would
develop wings, but was yet wingless. To prevent sexual contact, he isolated the aphid prior to molting. After developing
wings, it brought forth its progeny, thus proving that mating,
if it occurred, had taken place prior to the last transformation. This experiment was repeated several times to be sure
that no precaution had been neglected to establish the fact
beyond doubt.

The next step was to establish whether any coupling at all
was necessary, and Réaumur suggested that an experiment
could be easily designed to test this hypothesis. "This experiment is to observe a mother aphid give birth, and to take
care to bring up the new-born aphid in a place where it can
have no contact with other aphids."[62] Réaumur had tried this
experiment of isolating the offspring of a mother aphid immediately after birth several times but the longest that he
was able to keep an aphid alive was nine days. He suggested
that other naturalists might be more adept at nursing along
a solitary aphid.

Taking up Réaumur's challenge, Bonnet designed an experiment in which an isolated female aphid could be kept
under constant observation. He described to Réaumur the
precautions which he took to protect the virginity of his new-
born to prevent chance impregnation by an errant male.

During our past Spring vacation, I made several observations to
which I have not been indifferent, among others some on Aphids.
I attempted the experiment which you proposed for deciding
whether they have been granted the ability to multiply without

---

[61]"Nous n'avons donc trouvé jusqu'ici que les mères parmi les pucerons, nous n'y
avons point trouvé d'insectes que nous puissions regarder comme les mâles; les
deux sexes sont-ils réunis chés elles, comme ils le sont dans les limaçons? Il semble
que cela ne suffise pas encore, on voit les limaçons s'accoupler; & en quelque temps
que j'aye observé les pucerons soit aîlés, soit non-aîlés, je n'ai jamais apperçû aucun
accouplement." Réaumur, *Histoire des insectes* 3:327.
[62]Ibid., 329.

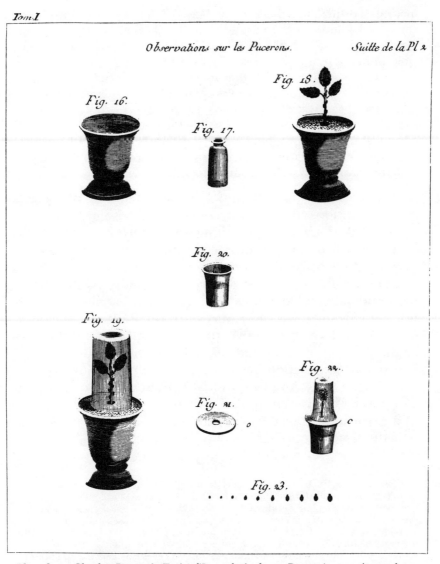

Plate from Charles Bonnet's *Traité d'Insectologie* shows Bonnet's experimental apparatus to isolate a female aphid immediately after birth to prevent impregnation. Successive virgin births established parthenogenesis.

(Permission to reproduce from Bonnet's *Oeuvres d'Histoire naturelle et de Philosophie* [Neuchâtel: Samuel Fauche, 1779] courtesy of Department of Special Collections, Case Western Reserve University Libraries.)

mating. To this end, the 20th of May, I filled a flower Vase with earth and I sunk in the middle of it a flask filled to its neck with water. In this flask I placed a little branch of the spindle tree on which I had left only five or six leaves, after having examined them attentively to assure myself that there were no aphids on them. I then placed on one of these leaves an aphid whose wingless mother, taken from the spindle tree, had just given birth under my very eyes. I then covered the little branch with a glass Vase which fitted exactly at its opening against the surface of the earth of the flower Vase, serving me better as a place for safe keeping than the Tower of Bronze where Danae was imprisoned.[63]

Bonnet patiently watched and recorded the births of successive generations of aphids produced by his isolated female. At the end of assiduous watching and record keeping which lasted over a month, Bonnet communicated by letter of 13 July 1740 his discovery to Réaumur. As he later recalled, "I put before him a table of the days and hours of the births by my female androgenous aphid, which I hardly let out of my sight from the 20th of May until the 24th of June, and for whom I have been an Argus more vigilant than the one of the fable."[64] Bonnet had proved that the aphid which he had placed in isolation had been able to produce ninety-five offspring. On 5 August 1740, Réaumur wrote a letter of congratulations to Bonnet. "These are assuredly observations of a great importance in natural history since they teach us that the law of coupling is not a general law."[65]

   Given the precision of Bonnet's experiments on the aphid

---

[63]"Pendant nos Vacances du Printems dernier j'ai fait quelques observations qui ne m'ont pas été indifferentes, entr'autres sur les *Pucerons.* J'ai tenté l'experience que vous proposés pour decider s'il leur est accordé de se multiplier sans accouplement. Pour cet effet, le 20ᵐᵉ May je remplis de terre un Vase à mettre des fleurs, au milieu duquel j'enfonçai jusqu'au col une fiole pleine d'eau. Je mis dans cette fiole une petite branche de Fusain à qui je ne laissai, que cinq à six feuilles, après les avoir examinées attentivement pour m'assûrer qu'elles n'avoient aucun Puceron. Je plaçai ensuite, sur une de ces feuilles, un Puceron dont la Mère dépourvue d'aîles et prise sur le Fusain, venoit d'accoucher sous mes yeux. Je couvris enfin la petite branche d'un Vase de Verre qui l'appliquant exactement par son ouverture contre la surface de la terre du vase à fleurs, me repondoit mieux de depôt que la Tour d'Airain ou Danae fut renfermée." Bonnet to Réaumur, 13 July 1740, Papers of Réaumur, Bonnet (DB), Archives de l'Académie des Sciences, Paris. The king of Argos imprisoned his daughter Danae in a tower because the Delphic oracle had predicted that his death would be at the hands of his daughter's son. It was to no avail. Zeus visited her in the guise of a shower of gold and she bore Perseus.
[64]Bonnet, *Mém. aut.,* 59.
[65]Ibid.

and his other observations on caterpillars, there is no doubt that Bonnet's interests during his student years focused on empirical studies of insects. Nevertheless, the development of more philosophical interests after 1745, in particular his interest in Leibniz, did not develop by chance within a philosophical vacuum.

The philosophical orientation of the curriculum in which his empirical investigations were undertaken was Newtonian. The Newtonian idea of providence provided the needed antidote to the excessively mechanical universe of Descartes. However, Leibnizian mathematics were held in high regard by Cramer and Calandrini. A careful examination of Trembley's thesis on the infinitestimal calculus under Calandrini to see whether he used the Leibnizian rather than the cumbersome Newtonian method of fluxions (which was not particularly popular on the Continent) is necessary to establish how influential Leibnizian mathematics was prior to 1740. If Cramer and Calandrini took heed of Leibnizian metaphysics before the discovery of the polyp, they do not appear to have communicated this interest to Trembley and Bonnet by encouraging them to read Leibnizian works.

Works by Leibniz are conspicuously absent in Bonnet's carefully guided list of recommended readings. This is not surprising because, though Cramer and Calandrini were undoubtedly familiar with Leibnizian mathematics through Cramer's close association with the Bernoulli circle in Basel, there is no indication that they were drawn to Leibnizian metaphysics at this time. Indeed, the period between 1720 and 1740, which included the years when Cramer and Calandrini directed the educations of Bonnet and Trembley, was the very time when interest in Leibnizian metaphysics, or any metaphysics for that matter, was eclipsed by the growing enthusiasm for observation and experiment associated in European minds with Newton.[66] The critical attitude toward Cartesianism which is apparent in the remarks made by Cramer in his *Cours de logique* (quoted above), was extended to all metaphysical speculation and Leibnizian philosophy fell victim to this widespread reaction against *a priori* truths, even

---

[66]W.H. Barber, *Leibniz in France: From Arnauld to Voltaire* (Oxford: Clarendon Press, 1955), 99.

though Leibniz, ironically, had been just as harsh a critic of Descartes as Newton himself.

Lack of interest in Leibnizian philosophy may also have been due to the scarcity of his works. Leibniz had seldom taken the trouble to gather his writings in book form. Much of his thought was scattered through a wide array of periodicals. The only substantial publications during his lifetime were the "Système nouveau de la nature et de la communication des substances" published in 1695 in the *Journal des Savants* and the *Essais de Théodicée,* published in 1710. Since the "Système nouveau" was published in a periodical, as the years passed, access to it must have become more difficult. Moreover, the publication of the *Théodicée* coincided with the decline of interest in works of a purely theological and metaphysical nature.[67] There was little critical response at the time of its publication. The most important work published before the 1740s which gave the French reading public access to the ideas of Leibniz appeared shortly after his death. This was his famous correspondence with Samuel Clarke which was first published in 1717. The French edition which was published in Amsterdam in 1720, edited by Des Maizeaux, *Recueil de diverses pièces sur la philosophie, la religion naturelle, l'histoire, les mathématiques, etc.* was owned by Charles Bonnet.[68] It is not known when he read it. Bonnet, who later became Leibniz's champion, admits in his *Mémoires autobiographiques* that he had some acquaintance of Leibniz through Fontenelle's *Éloge* published in the *Histoire de l'Académie Royale des Sciences.*[69] In 1748, after the failure of his eyesight, he read the *Théodicée* for the first time. This work, he recalled, was a revelation: "The Theodicy was for my mind a sort of telescope, which uncovered for me another universe, whose view appeared to me as an enchanted, I would say almost magical, perspective."[70]

However, Trembley, it appears, did not try to read the

---

[67]Ibid., 92.

[68]This edition published by Sauzet exists in the George Trembley Archives with Charles Bonnet's signature on the title page.

[69]Charles Bonnet, *Mém. aut.,* 100. See also *Histoire de l'Académie Royale des Sciences,* 1716 (Paris: Pankoucke, 1777), "Eloge de M. Leibnitz," 115–56.

[70]"La Théodicée fut pour mon esprit une espèce de télescope, qui me découvrit un autre univers, dont la vue me parut une perspective enchantée, je dirais presque magique." Charles Bonnet, *Mém. aut.,* 100.

Théodicée until late in his life. In 1784, the year of Trembley's death, he was struggling to understand the work and Bonnet sympathetically offered the index which he had prepared many years before.[71] Bonnet wrote to Trembley:

The *Théodicée* is a beautiful forest, where the paths have not been cleared and where sometimes the underbrush scratches the legs of the traveller . . . Some admirable things can be found there mixed with others which are not. The author collected there all his metaphysical, theological and moral principles.[72]

Trembley's contact with the empirical Newtonian-Baconian circle at Leiden which played a direct role in the discovery of the polyp, can be viewed as a continuation of the Newtonian orientation of the Academy. In particular, as the following chapter will demonstrate, the correspondence between Réaumur and Trembley shows the important influence of Boerhaave on Trembley's conceptualization of the polyp's plant-animal nature and on the particular experiments he devised to explore the problem. The unfolding of the discovery of the polyp through the Trembley-Réaumur letters demonstrates not Réaumur's influence, but rather their differences in interpretation of the discovery.

---

[71]Bonnet's index on the flyleaf of his copy of *Essàis de Théodicée sur la bonté de Dieu, la liberté de l'homme, et l'origine du mal* (Amsterdam: David Mortier, 1720) in the George Trembley Archives may be the one referred to here by Bonnet.

[72]Bonnet to Trembley, 30 March 1784, George Trembley Archives, Toronto, Ontario.

Chapter 4

The Discovery of the Polyp

Trembley's life at Sorgvliet, Count Bentinck's country estate, appears to have allowed him considerable leisure. Removed from the stimulating intellectual atmosphere of Leiden and far from the political troubles at home and from family demands, Trembley's limited duties as tutor to the nobleman's two young sons, Jean and Antoine, allowed him time to wander over the dunes which stretched to the sea behind the Bentinck estate in search of unusual insects. The vignettes which introduce each of Trembley's four *Mémoires* show him with his pupils at various locations on the estate. It is clear that the children participated in Trembley's exploration of Nature's artistry. The Bentinck estate became an extension of the classroom. The formal gardens, fishpond and artificial ditches that enhanced the classic balance of the mansion were carefully searched by Trembley and his students for terrestrial and aquatic insects. Trembley allowed

his students to scramble up on the chairs in his study to peer
into the jars on his window sills in which he kept their specimens. In Trembley's mind, the study of natural history was
admirably suited to the formation of intellectual and moral
qualities in the young. To appreciate the variety and intricacy
of Nature's designs was the better to admire the Creator.
Rather than a frivolous pastime, the study of natural history
was both serious and difficult.

Generally, a hand-lens or a simple microscope was all the
apparatus that Trembley required to make his careful observations of insects.[1] There is little evidence that he used
the unreliable compound microscope of the period with its
high degree of chromatic and spherical aberration. Often he
made his studies of aquatic insects with a magnifying glass
in the natural light from two large windows in his room
looking out on the mansion's decorative pond and its semi-
circular greenhouse. For objects that demanded a higher
degree of magnification, Trembley used a simple microscope,
probably of his own design and constructed by one of the
reputable Leiden instrument makers, schooled in the difficult
art of lens grinding in which Leeuwenhoek had excelled.

Trembley's instrument maker mounted a jointed support
arm, consisting of three "Musschenbroek nuts," on a station-
ary platform. To this moveable arm, he attached a single
lens. Trembley could place a jar containing his specimens on
the platform and position the lens at a desired point without
interfering with the activities of the pond life within the jar.[1]
After darkening the room, he placed a candle, whose flame
was exactly the height of the object, behind the jar. Squinting
with his eye against the lens, he could take advantage of the
considerable magnification of the lens, as well as that produced by the refraction of the light through the water. No
doubt Trembley was able to afford this elegant and certainly
costly instrument because of the generosity and keen interest
in his employer and patron, William Bentinck.

Trembley's vocation as a tutor at Sorgvliet was not prac-

---

[1]See picture and description of this apparatus in *Nouvelles découvertes faites avec le
microscope, par T. Needham traduites de l'anglois, avec un mémoire sur les polypes à bouquet
et sur ceux en entonnoir*, par A. Trembley (Leyden: E. Luzac, 1747). The translator
notes in the preface that Trembley's "microscope" is available in Leiden from Musschenbroek. See also John R. Baker, *Abraham Trembley*, 172–73.

ticed in complete isolation from the stimulation of other minds. As Margaret Jacob has shown, the Bentinck family played an important role in the political and intellectual life of the Netherlands in the early eighteenth century.[2] The vigorous social life of William and Charles Bentinck included a variety of intellectuals, whose interests encompassed a spectrum from the more conservative, yet controversial, Newtonian providentialism of 'sGravesande to radical republican ideas and pantheistic doctrines of Free Masonry, rooted in the secret practices of Renaissance hermeticism. Moreover, Trembley's relative solitude on William Bentinck's country estate was balanced by his close proximity to The Hague, only about a mile and a half away. A city of lively coffee houses, the city had attracted many Protestant *emigrés*, driven from France after the revocation of the Edict of Nantes. The literate and heterogeneous population supported a flourishing book trade.

It is likely that if the literary club to which the now aging 'sGravesande had belonged in the heyday of the publication of his *Journal Littéraire* still met there in the 1730s that Trembley would have been invited, for he went to The Hague frequently to visit his friend Pierre Lyonet. Of French-Swiss Calvinist background, Lyonet worked for the government as a translator and expert on the decoding of secret documents. What drew Trembley and Lyonet together was their passion for the study of insects, inspired by Swammerdam and Réaumur's *Histoire des insectes*. At this time Lyonet was also corresponding with Réaumur. An artist of considerable talent, the illustrations for Trembley's *Mémoires* came from his hand. He drew with the trained eye of a naturalist. As Trembley wrote in praise of Lyonet: "It was sufficient for me to put objects before Monsieur Lyonet's eyes, in order for him to see everything that would have been difficult to indicate to others. Not only was he a skilled draughtsman, but also a keen and experienced observer."[3] In addition to his work as

---

[2]See Margaret Jacob, *The Radical Enlightenment: Pantheists, Freemasons and Republicans* (London: George Allen and Unwin, 1981).

[3]"Il m'a suffi de mettre les objets sous les yeux de Mr. Lyonet, pour qu'il vît tout ce qu'il auroit été difficile de faire remarquer à d'autres. C'est qu'il est non seulement habile Dessinateur, mais encore Observateur pénétrant & experimenté." Trembley, *Mémoires, pour servir à l'histoire d'un genre de polypes d'eau douce, à bras en forme de cornes*

Trembley's illustrator, Lyonet engraved eight of the copper plates to the Leiden edition (numbers 8 through 13), after only cursory instruction by the Dutch master engraver, Jan Wandelaar (1690–1759).

Like Bonnet, Lyonet had begun to consider the possibility that aphids might reproduce without mating after reading Réaumur's invitation in Volume Three of the *Histoire des insectes*. In fact, he claimed to have discovered parthenogenesis independently of Bonnet. After Bonnet's experiments established the unusual mode of reproduction in aphids, Lyonet was among those selected by Réaumur to confirm the discovery. Lyonet did not publish his observations on insects separately in the early years of this association. However, the copious notes to his translation of Lesser's *Théologie des insectes ou démonstration des perfections de Dieu dans tout ce qui concerne les insectes* contain some of his own meticulous observations that expand upon and correct Lesser's rather mediocre work. In addition, Trembley's letters to Bonnet during the years that he resided at Sorgvliet show Lyonet's intense involvement with the perplexing issues of parthenogenesis and regeneration.

I. THE INFLUENCE OF BONNET'S DISCOVERY OF PARTHENOGENESIS

Since Réaumur had also asked Trembley to confirm Bonnet's discovery of the ability of aphids to reproduce without mating, Trembley selected a species of aphid that feeds on the leaves of the elder. He was successful in producing several generations of aphids from a single isolated mother in the fall of 1740. However, Trembley was not entirely convinced that all the questions about aphid reproduction had been satisfactorily answered and encouraged Bonnet to continue his study of aphid reproduction. Trembley's reservations may have been due to his association and collaboration with Ly-

(Leiden: Verbeek, 1744), préface, vi. The Paris publisher Durand printed a second edition in 2 volumes the same year without Trembley's approval. For extensive biographical and archival source information on Lyonet see: W. H. Van Seters, *Pierre Lyonet, 1706–1789* (La Haye: Martinus Nijhoff, 1962).

onet. Did Bonnet's experiments prove that no coupling what-
soever took place?

As Bonnet, Trembley and Lyonet carefully studied different species of aphids, they became able to identify both male
and female, something that Réaumur had failed to do. Bonnet noted that the smaller male was "perhaps one of the most
ardent that there is in Nature. It appears to me that it does
nothing except have intercourse as soon as the day arrives."[4]
Thus, when circumstances permitted, coupling did take
place! As Trembley's letter of 27 January 1741 shows, he was
not entirely satisfied that Bonnet had proved true virgin
birth. The discovery of the male made him question whether
no coupling at all had taken place during Bonnet's earlier
experiments. Rather than an exception to natural law, Trembley proposed two possible alternative explanations of the
facts. The first was one which Réaumur himself had considered: male and female, upon reaching a certain degree of
maturity *in utero* might mate prior to birth. In this case, Trembley reasoned, perhaps the aphids would be born in pairs
like twins at short time intervals apart. His own experiments,
as one might have suspected, were inconclusive. However,
he suggested to Bonnet that in such extraordinary circumstances, circumstances that challenged the existing ideas of
animal reproduction, no approach should be neglected.

The second possibility Trembley proposed as a question:
"In cases so far from ordinary circumstances, trying everything is permitted. I said to myself: 'Who knows if one mating
might not serve for several generations?' "[5] Lyonet, he reported in his letter, claimed to have seen mating in another
species of aphid. The males were winged and smaller than
their female counterparts. Though Trembley attested to his
great confidence in Lyonet as an observer, he emphasized
the need to repeat these experiments "with great precautions
before being convinced of anything."[6] What he did not report
to Bonnet was that Lyonet had concluded that one coupling

---

[4]Bonnet to Trembley, 18 December 1740, George Trembley Archives, Toronto,
Ontario. (For text of letter, see Appendix A below, 196.)

[5]Trembley to Bonnet, 27 January 1741, Ms Bonnet 24, Bibliothèque Publique et
Universitaire de Genève. (See below, 198.)

[6]Ibid.

did indeed suffice for several generations.[7] This was, of course, a direct challenge to Bonnet's claim to have established true virgin birth in his aphids.

Bonnet dismissed the idea that aphids were able to mate within the belly of the mother. The physical state of the embryos in the womb prevented it. Moreover, he was not sure that Réaumur himself accepted this explanation, even though he had been the one to first propose it. He wrote to Trembley:

This conjecture does not agree with the state of the Aphids enclosed in the womb, where they are not only washed in a fluid which does not permit them to unite, but also where they are enclosed by a membrane which keeps all their parts better bound up than those of chrysalises. Unless you desire that the coupling of Aphids happens as Swammerdam imagined that of Bees, or in some other analogous manner, I do not see that it is conceivable that it is carried out in the Bosom of the Mother.[8]

Trembley's second suggestion (and the view that Lyonet supported), that one mating might serve to produce several generations of aphids, Bonnet found worthy of serious consideration. "With respect to yours [your idea], it pleases me a great deal and assuredly merits study," he wrote.[9]

Trembley's query stimulated Bonnet to undertake the most arduous observations of his career. Experimenting on several species of aphids, he succeeded in bringing up nine successive generations. A representative of each generation was selected at birth and kept in isolation until it produced offspring. Its offspring were then each placed in separate containers until Bonnet had proved that nine generations of lice were produced without the amorous attendance of a single male!

This experimental program had unfortunate consequences. Driven by the question in Trembley's letter, Bonnet ruined his eyesight, possibly from the strain of keeping an hourly vigil of his plant lice day and night for a period of

---

[7]Lesser, *Theologie des insectes . . . avec des remarques de M.P. Lyonnet*, (Paris: Chaubert & Durand, 1745), 57.

[8]Bonnet to Trembley, 24 March 1741, George Trembley Archives, Toronto, Ontario. (See below, 200.) Réaumur had suggested that aphids might couple in the womb in *Mémoires des insectes* 3: 329. Swammerdam believed that male bees did not copulate with the female, but fertilized the eggs by ejecting their sperm "in the same manner of Fishes, who only shed it upon the spawn." See *The Book of Nature*, tr. Thomas Flloyd, reprint edition (N.Y.: Arno Press, 1978), Part I, 187, column b.

[9]Ibid. (See below, 201.)

three months. As Bonnet later described the effect of Trembley's letter on him: "If this excellent friend had been able to foresee all the evil that his 'who knows' did to my eyes, I am very sure that his tender friendship for me would not have permitted him to express it. It was however, on this simple 'who knows' that I undertook a new study which was much more laborious than the preceding one. I was young and full of ardor: it seemed that these two words reduced to nothing all my previous work."[10]

In the light of twentieth-century embryology which has given us a greater understanding of parthenogenetic cell division, Trembley's question seems to have little significance. Nevertheless, it affords a glimpse into the metaphysical issues that surrounded the generation of animals prior to 1740. Parthenogenesis supported the ovist position that the miniature preexisted within the egg of the female. Indeed, Réaumur, in suggesting the problem to other naturalists, merely demanded a more rigorous empirical confirmation of a conclusion which had already been reached by Cestoni, Vallisnieri and Bourguet, all of whom were ovists.[11] Bonnet's successful proof of parthenogenesis, based on a rigorous empirical methodology, confirmed what had been expected. For ovists who believed that the egg carried the encapsulated miniature prior to fertilization, Trembley's question was merely one of determining more closely the mechanisms on which the development of the egg depended. Though vastly subordinate to the function of the egg, the sperm obviously played a role in generation, since in the observation of higher animals, the offspring had both male and female characteristics. Thus, it was not unreasonable to suggest that a coupling every few generations might still be necessary to ensure the continued unfolding of the preordained genealogy of plant lice. At first Trembley was reluctant to admit for aphids an exception to a natural law which held true for all other species

---

[10]"Si cet excellent ami avait pu prévoir tout le mal que son *qui sait* ferait à mes yeux, je suis bien sûr que sa tendre amitié pour moi ne lui aurait pas permis de le lâcher. Ce fut pourtant sur ce simple *qui sait* que j'entrepris un nouveau travail beaucoup plus pénible que le précédent. J'etais jeune et plein d'ardeur; il me sembla que ces deux mots mettaient à néant tout ce que j'avais fait." Bonnet, *Mém. aut.*, 63.

[11]Louis Bourguet, *Lettres philosophiques*, (Amsterdam: Marc-Michel Rey, 2nd ed., 1762), 96.

of animals: even snails and other hermaphrodites needed both male and female parts to reproduce.

The proof that aphids could reproduce without coupling was to have a decisive influence on Trembley's discovery of the polyp. It made him wary of unproven generalizations about the reproduction of animals. Trembley wrote: "A fact such as the one which aphids presented could only inspire in me a great deal of distrust of general laws . . . I felt strongly that Nature was too vast, and too little known, for anyone to decide without foolhardiness that one or another property was not found in such and such class of organized bodies."[12] Likewise, Réaumur warned his readers that, as a consequence of Bonnet's discovery, all assumptions, as far as animal reproduction was concerned, were suspect. Before Bonnet's discovery, "Everything had pointed to the necessity of the coming together of two individuals of the same species for propagation of every species. It was believed that the Author of Nature intended that this law be a general one."[13] Hermaphroditic animals still required the joining of male and female and thus could still be assumed to reproduce according to Réaumur's first rule, as set forth in Volume Two of his *Histoire des insectes:* all animals require the coupling of male and female and bring forth either eggs if they are oviparous, or living young of they are viviparous.[14] This wariness of general laws shows how far Trembley and his contemporaries had come from the complacent Cartesian assumption that all of nature would eventually be comprehended under a few simple laws. God's design was infinitely more complex and possibly never fully comprehensible.

The careful study of aphids had demonstrated that no law was incontestable. In addition to establishing parthenogen-

---

[12]"Un Fait tel que celui que présentoient ces Pucerons, ne pouvoit que m'inspirer beaucoup de défiance pour les régles générales. . . . Je sentois vivement, que la Nature étoit trop vaste, & trop peu connuë pour qu'on pût décider sans témerité, que telle ou telle propriété ne se trouvoit pas dans telle ou telle classe de corps organisés." Trembley, *Mémoires*, 18.

[13]"Jusqu'ici tout a prouvé la nécessité du concours de deux individus de la même espece pour la propagation de chaque espèce. On croyoit que l'Auteur de la Nature avoit voulu que cette loi fût générale." Réaumur, *Histoire des insectes* 6: préface, xlvi.

[14]"Qu'il n'y a aucune espèce d'insecte qui ne soit ovipare ou vivipare; qui ne se perpétue, soit en pendant des oeufs, soit en mettant au jour des petits vivans." Réaumur, *Histoire des insectes* 2: xxxix.

esis, the observations of Bonnet and Trembley, but particularly those of Lyonet, demonstrated another exception to one of Réaumur's cherished dicta: animals were either viviparous *or* oviparous. Lyonet demonstrated that aphids were capable of producing living young and eggs, depending on the season of the year.

In the course of their examination of aphid reproduction, Bonnet, Trembley and Lyonet observed that in late autumn the female produced, not living young, but little oblong bodies. As early as 18 December 1740, Bonnet wrote to Trembley, "The females also offer the singularity of producing at one time living aphids, at another time fetuses that they arrange one beside the other, the way Butterflies arrange their eggs."[15] Trembley responded, 27 January 1741: "What you tell me about fetuses which aphids produce is also known to us here."[16] In his letter to Bonnet, dated 23 August 1742, Trembley described more fully Lyonet's significant observations which established his priority to the discovery.

Last April we took a walk together in the woods of Sorgvliet and Mr. Lyonnet [sic], who sees all, discovered on the bark of an oak some little oblong and brownish bodies which greatly resembled what you regard as the aborted fetuses of aphids and which he believes are eggs. At once he was of the opinion that these bodies were eggs, and he took some to his study, from which he saw aphids emerge, which belong to the large species described in Mr. de Réaumur's 3rd volume.[17]

In 1742 when Réaumur published Volume Six of his *Histoire des Insectes,* he gave Bonnet, rather than Lyonet, the credit for suggesting that the small oblong bodies that aphids produced in the autumn were eggs. At the same time he admitted that he did not subscribe to this view. Instead, he suggested that they were immature fetuses that the mother had to abort before the advent of the winter. Describing Bonnet's observation that it was only those aphids which Bonnet had observed to mate with the male that produced the

[15]Bonnet to Trembley, 18 December 1740, George Trembley Archives, Toronto, Ontario. (See below, 196.)

[16]Trembley to Bonnet, 27 January 1741, Ms Bonnet 24, Bibliothèque Publique et Universitaire de Genève. (See below, 197.)

[17]Trembley to Bonnet, 23 August 1742, Ms Bonnet 24, Bibliothèque Publique et Universitaire de Genève. (See below, 213–14.)

aborted fetuses, he concluded that it appeared that the only
function of intercourse in aphids was to rid the mothers of
those fetuses which had not reached full term.[18]

In his letter of 5 September 1743 to Trembley, Bonnet
returned to the subject of Réaumur's aborted fetuses, claim-
ing, it would seem without justification, that he had never
thought that they were anything but eggs:

Read, please from page 558 of the 6th volume of the *Mémoires sur
des insectes,* up to page 560 and you will see that Mr. de Réaumur
considers these little oblong bodies to which aphids sometimes give
birth to be aborted fetuses, but which I, myself, have always had
the inclination to suspect were eggs. It is then with very palpable
pleasure that I see my conjecture has been verified.[19]

It appears that Trembley, to forestall Bonnet from receiving undue credit for the idea that aphids produce eggs, reported in the preface to his *Mémoires* on the polyp that Lyonet
had established that the species of aphids which lives on the
bark of the oak was viviparous in the summer, but toward
the end of autumn, the last generation produced, not living
young, but eggs. In the spring these eggs hatched to begin
the reproductive cycle again.[20]

Do the experiments on aphids by Bonnet and Trembley
demonstrate that they began their experimental work as convinced ovists? Did the microscope and magnifying glass become the instruments of those determined to prove a pre-
conceived idea of the supremacy of the egg over sperm? Here
a study of their letters and published works demands respect
for the delicate relationship between theory and observation
which comprises the process of discovery. For both Bonnet
and Trembley, the facts of observation were to be weighed
against the general rules. The only truths were those revealed
through a study of nature's objects. While there is no doubt
that some of the works that Bonnet read during his education
favored the ovist interpretation, and the association of Cramer and Calandrini with Bourguet may have confirmed this
tendency, hard empirical fact, not speculation, was the aim

---

[18]Réaumur, *Histoire des insectes* 6: 559.

[19]Bonnet to Trembley, 5 September 1743, George Trembley Archives, Toronto,
Ontario. (See below, 227.)

[20]A. Trembley, *Mémoires*, préface, ix–x.

of the natural history which Bonnet and Trembley pursued with such passion and exactitude during this period. This undoubtedly reflects the influence of the observational methodology of Réaumur.

Far from settling the questions which divided the ovists and the animalculists, parthenogenesis revealed the difficulty of explaining the mechanisms which governed reproduction. The discovery reinforced Trembley's determination to be extremely circumspect when it came to concluding anything. For this reason the discovery of parthenogenesis must be seen as extraordinarily important in the subsequent discovery of the polyp. The scrupulous attention which he gave to every aspect of the polyp's morphology was an effect of the unsettled questions about animal generation that the reproduction of aphids had raised.

It should be emphasized that a study of Trembley's letters during this period, particularly his letters to Bonnet, reveals that Trembley was intensely involved in all the perplexing issues surrounding the reproduction of aphids at the same time that he was beginning to explore the peculiar characteristics of the polyp. In both the Trembley-Réaumur and the Trembley-Bonnet correspondence, parthenogenesis and the polyp were often discussed in the same letters. The unsettling questions presented by the reproduction of aphids, more than any other factor, made Trembley determined to avoid precipitous conclusions about insect reproduction.

As the discovery of the polyp unfolded, the letters which Trembley and Réaumur exchanged show, not the similarities in point of view between Réaumur, the supposed teacher, and Trembley, the disciple, but rather their differences in interpretation. In the letters it is Réaumur who is dumbfounded. He is the student who must reluctantly accept the facts once he has seen them with his own eyes. Describing the effect that Trembley's discovery of regeneration had on him in his *Histoire des insectes*, Réaumur wrote: "Still I confess that when I saw for the first time two polyps form little by little from the one I had cut in two, I could hardly believe my eyes; and it is a fact which I am not at all accustomed to seeing, after having seen it over again hundreds and hundreds of times."[21]

---

[21]"J'avouë pourtant que lorsque je vis pour la premiére fois deux polypes se former

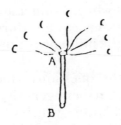

Trembley's first sketch of the polyp in a letter to Réaumur, 15 December 1740. Courtesy Georg et Cie, Librairie de l'Université, Genève. From Maurice Trembley, *Correspondance inédite entre Réaumer et Abraham Trembley* (1943).

II. THE TREMBLEY-RÉAUMUR LETTERS ON THE POLYP

Trembley communicated for the first time his perplexing observations of the curious organism that Réaumur later named the polyp in his letter of 15 December 1740. The previous June he had noticed in the beaker of water in which he kept various aquatic insects "something" attached to the sides of the glass and on the stems of the aquatic plants. "It is green, and seems at first glance to be a plant. Then one discovers several characteristics of an animal." When touched or the beaker shaken, the body suddenly contracted, but when at rest it resembled an extended corridor fixed at one end and surrounded at the other end by threads which were attached perpendicularly to the body. "These threads can move themselves in all senses independently of each other, and even when the body remains tranquil."[22] He described the polyp's ability to move about, its light seeking tendencies, and finally its ability, after being cut in half, to form two separate living entities. Trembley performed this crucial experiment on 25 November 1740.

I first thought of the feet and the antennae of crayfish which grow back; but the difference is that the two portions . . . seem actually two complete animals; in such a manner that one could say that of one animal, two have been produced. There is much resemblance

peu à peu de celui que j'avois coupé en deux, j'eus peine à en croire mes yeux; & c'est un fait que je n'accoutûme point à voir, après l'avoir vû & revû cent & cent fois." Réaumur, *Histoire des insectes* 6: liv–lv.

[22]Maurice Trembley, *Correspondance inédite*, 11.

between what happens to this animal, and plants which grow from cuttings. Perhaps the threads c. c. c. fig. 1 are sorts of roots and that what I call an animal is perhaps an ambulant and sensitive plant, but which has characteristics very different from all those which one knows.[23]

It should be noted that though Trembley had been carefully observing the phenomenon for six months, the crucial experiment, that of cutting it, had only been made the previous month. The reason for the experiment was Trembley's question whether the "thing" was animal or plant.

In his *Mémoires* Trembley revealed the steps in his thinking that led up to the discovery that polyps could regenerate from cuttings. The first polyps which he had collected had been scooped up along with a number of aquatic insects from a ditch at Sorgvliet. He had placed them all in a large jar on the window sill of his study. At first his attention was attracted to the more active insects in the jar. The shape and green color of his first polyps, as well as their apparent immobility, led him to believe that they were a type of parasitic plant, since he observed them affixed to some of the aquatic plants in his jar.[24] The first characteristic of the polyp which excited his curiosity was the motion of the tentacles.

Thinking that the Polyps were plants, I could hardly imagine that the movement of these slender threads, located at one end of their bodies, was their own. Yet they appeared to move by themselves and not at all as a response to the agitation of the water . . . Subsequently, the more I studied the movement of the arms, the more it appeared that it had to come from an internal cause, and not from an impetus external to the Polyps.[25]

---

[23]"J'ai d'abord pensé aux pieds et aux antennes des écrevisses qui recroissent; mais la différence qu'il y a, c'est que les deux parties . . . paraissent actuellement deux animaux complets; en sorte que l'on pourrait dire que d'un animal on en a fait deux. Il y a beaucoup de ressemblance entre ce qui est arrivé à cet animal, et les plantes qui viennent de bouture. Peut-être les fils c. c. c. fig. 1. sont-ils des espèces de racines et ce que j'appelle un animal est peut-être une plante ambulante et sensitive, mais qui aurait des caractères bien différens de toutes celles que l'on connait." Ibid., 14.

[24]A. Trembley, *Mémoires*, 7–8.

[25]"Dans l'idée que j'avois que les Polypes étoient des Plantes, je ne pouvois guéres penser que ce mouvement, que j'observois dans ces fils déliés qu'ils avoient a une de leurs extremités, leur fut propre; & cependant il paroissoit tel, & nullement l'effet de l'agitation de l'eau . . . plus je considerai dans la suite le mouvement de ces bras, plus il me parut devoir venir d'une cause intérieure, & non d'une impulsion étrangère aux Polypes." Ibid., 9.

When he shook the jar gently, to his surprise he saw the bodies of the polyps immediately contract, so that they then appeared as no more than specks of green matter in his jar. As he studied these specks with a magnifying glass, he saw them begin to stretch out again. The independent motion of the tentacles and the contractility of their bodies gave him the strong impression that they were animals. Not long afterwards Trembley discovered the polyps' mode of locomotion—a type of head-over-heels walk, similar to that of inch worms. This convinced Trembley that they were animals and he gave up studying them.

However, late in September 1740, they again piqued his curiosity. He noticed that his polyps seemed to cluster on the side of the jar facing the sunlight. Experiments confirmed that they had a "definite propensity for the best lit area of the jar."[26] Without concluding anything at this point, he resolved to make the natural history of the polyp an object of an extensive study.

Not long afterwards, he noticed that not all of his polyps had an equal number of tentacles and, comparing them to the unequal number of branches and roots characteristic of plants, he again suspected that they might be plants. This gave him the idea of cutting them. "I thought that if the two parts of the same Polyp lived after having been separated and each became a perfect Polyp, it would be clear that these organized bodies were Plants. Nevertheless, as I was much more inclined to believe that they were Animals, I did not set much store by this experiment; I expected to see the cut polyps die."[27]

In the first letter to Réaumur in which he discussed his discovery, Trembley confessed to Réaumur that he did not know what to make of it. Had it already been discovered? In the rough draft of his letter to Réaumur, he added, "I will continue as if it were not."[28] Trembley's ignorance of the

---

[26]Ibid., 12.

[27]"Je jugeai que, si les deux parties d'une même Polype vivoient après avoir été séparées, & devenoient chacune un Polype parfait, il seroit évident que ces corps organisés étoient des Plantes. Comme cependant j'étois beaucoup plus porté à croire que c'étoient des Animaux, je ne comptois pas beaucoup sur cette Expérience; je m'attendois à voir mourir ces Polypes coupés." Ibid., 13.

[28]Maurice Trembley, *Correspondance inédite*, 14, note 2.

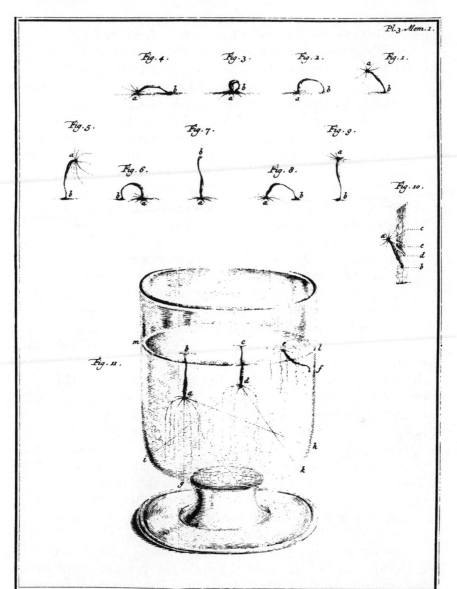

Plate from A. Trembley, *Mémoires*, shows the various modes of locomotion employed by the polyp, as well as its ability to anchor itself within a powder jar.
fig. 1–4: characteristic inch-worm-like gait
fig. 5–9: head-over-heels motion
fig. 10: progressive motion of body column using arm as anchor
fig. 11: two polyps suspended from the surface of the water using four arms fixed to the bottom as cables or anchors to moor themselves. A third polyp in the process of detaching from the side of the jar to assume the surface suspension.

prior discovery of the polyp by Leeuwenhoek and an anonymous Englishman, reported in the *Philosophical Transactions
of the Royal Society* for the year 1703, was fortuitous.[29] Both
had described it as an animal, and Leeuwenhoek had called
attention to its reproduction through budding. However, neither one had thought to cut the polyp, so that its ability to
regenerate escaped notice. The discovery was lost amid the
reports of other, often trivial, observations that crowded the
pages of the *Philosophical Transactions* of the early eighteenth
century. Unaware of these reports, as well as the observations
of Bernard de Jussieu who had described it as a plant, Trembley continued to mull over the question of whether his little
creature was animal or plant.

The ability of the creature to regenerate from cuttings,
Trembley considered an artificial form of reproduction because it was induced by sectioning. Trembley's quest to discover how it naturally reproduced drove him to an ever more
careful and demanding experimental program. Indeed, at
first he believed that an understanding of the polyp's natural
mode of reproduction would provide the key to unraveling
the mysteries of its structure. "If I arrive at discovering this,"
he wrote in a letter to Réaumur, "I could more easily judge
by the characteristics which it furnishes me the nature of this
being and that of its animated and separated parts."[30]

Réaumur, in his response to Trembley 15 January, reported that his curiosity had been extremely aroused by
Trembley's account of "these little organized bodies." By using the term "organized bodies" Réaumur was able to avoid
stating any definite conclusion with regard to its animal or
plant nature. He said, referring to Trembley's experiment
of cutting it in two: "Your last observations seem, however,
to decide that they are plants. But do plants walk? It is easier
for aquatic plants to move about than for terrestrial plants,
and plants in water can move towards the place which has
the most light or heat in the way that terrestrial plants conduct
themselves toward the open air, or those in a cave reach
toward the air hole. But what bothers me is that yours walk

---

[29]*Philosophical Transactions of the Royal Society*, (N.Y.: Johnson Repr. Corp., 1963)
23 (1702–03): 1304.
[30]Maurice Trembley, *Correspondance inédite*, 15.

on the glass."[31] The mobility of Trembley's little beings threw
Réaumur into a quandary.

On 16 February 1741, Trembley wrote another long letter
to Réaumur. Referring to the "enigma," Trembley said that
it was becoming deeper rather than lifting. "I am all the more
circumspect because the facts which the little bodies present
are new. I suspend my judgment. I do not dare to give it a
name, although this would be very convenient. That of *animal
plant* or *plant animal* presents itself rather naturally."[32] Greatly
perplexed, Trembley continued to juxtapose the plant and
animal characteristics in his mind. While the mobility of the
polyp made him lean to classification as an animal, other
plant-like characteristics raised certain doubts. In particular,
he noticed the uneven number of tentacles, mentioned above.
This irregularity was more characteristic of a plant.

During the months of July, August and September, I was nearly
continually persuaded that these little bodies were animals. I
founded this persuasion on the movements which I saw them make.
The air of a plant which they have when they are stretched and
above all some other characteristics threw me however from time
to time in doubt. The thought came to me to cut them. I regarded
this expedient as sufficient to decide the question. It furnished me
with characteristics which until now have been known only in plants.
However, I could not resolve my doubt. The characteristics of an
animal which these little bodies have held me back.[33]

When he cut a polyp in half and watched the regrowth of
each part into a complete body, Trembley began to lean
toward the plant explanation.

But at the same time I saw these parts walk, take steps, mount,
descend, shorten, lengthen, and make so many movements, which
up to this time I had only seen made by animals. I did not know

---

[31]Ibid., 17.

[32]Ibid., 23. The underlined words are Trembley's. This letter was read by Réaumur to the Academy of Sciences on March 1st, 8th and 22nd.

[33]"Pendant les mois de juillet, août et septembre, j'ai presque continuellement été
persuadé, que ces petits corps étaient des animaux. Je fondais cette persuasion sur
les mouvemens que je leur voyais faire. L'air de plante qu'ils ont, lorsqu'ils sont
allongés, et principalement d'autres caractères, me jetaient pourtant, de tems en
tems dans le doute. La pensée me vint d'en couper. Je regardai cet expédient comme
propre à décider la question. Il m'a fourni des caractères qui jusqu'ici n'ont été
connus que dans les plantes. Cependant je n'ai pu être tiré de mon doute. Les
caractères d'animal qu'ont ces petits corps m'y retiennent." Ibid., 24.

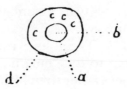

Cross section of the polyp, consisting of inner and outer membranes, *a* and *d*, and central cavity, *b*, from letter of 16 February 1741. Courtesy Georg et Cie, Librairie de l'Université, Genève. From Trembley, *Correspondance inédite*.

what to think. My curiosity was more and more piqued by that, and I resolved to try to unravel this enigma, by applying myself to discover the structure of these little organized bodies.[34]

But research on the structure of these bodies also failed to reveal characteristics which could clearly indicate whether the creature belonged to the animal, or vegetable, realm. By dissection Trembley found that these "little machines" consisted of two cylinders: one inside of the other. Normally, the central cylinder was empty like that of certain reed-like plants, but what puzzled him was that it was sometimes filled.[35] This was not plant-like. Between the two cylinder walls Trembley found grains of a green matter which appeared to be suspended in a sort of transparent liquid. "After having thus examined this green color, I wished to know how it was contained in the body; if it was a liquid analogous to the blood of animals or to the sap of plants."[36] Trembley concluded that the green matter was not contained in "vessels," but was spread throughout the body like "curds." Moreover, the green matter was only slightly viscous and contained a larger amount of solid matter than liquid.

The coarseness of the green matter and the bag in which it was contained showed us nothing analogous to the vessels and the blood of animals or to those of plants and to the sap which flows inside them. Rather, this bag was more like a stomach or intestine and the green matter like the food or excrement. But it is impossible to decide anything more. Are these green grains the seeds or the eggs of these little bodies?—another idea which comes to mind. It is necessary to prepare oneself for everything in these cases which

---

[34]Ibid., 28.
[35]Ibid., 31.
[36]Ibid., 32.

are so extraordinary . . . One is well acquainted with aquatic insects which are transparent and whose eggs can be seen through the body to give them their color. But the comparison is not very exact.[37]

Finally, he observed that the transparent skin is sticky like that of a snail and "if one abandons oneself to the penchant of judging by comparison, it is very tempting to conclude that these little bodies are animals. But can an animal cut in three pieces become three separate animals? This is what convinces me to suspend my judgment."[38]

The next mystery which Trembley tackled was why the inner cylinder sometimes appeared to be filled. Trembley observed the inner cylinder scrupulously, thinking at first that perhaps when the cylinder appeared filled, the inner surface had merely been pulled out, rather like that of a glove, and that this protrusion was possibly to assist in the locomotion of the creature. However, this did not seem to be a sufficiently likely explanation because he had viewed the phenomenon even when the body was at rest.

Passing from this unexplained fact, Trembley explored the question of the function of the threads which surrounded the opening at the anterior end. He noted again that each thread could move independently of the others. The movement of the threads was not a simple response to the agitation of the water, since they could move independently even when the water was calm. "I had many occasions to see that the principle of the movements of the threads was in the machine."[39] The "little machine" was capable of purposive activity. It was not merely pushed about aimlessly by the currents in the water. Trembley then reasoned that the threads assisted the little bodies in walking both on the surface of the glass and under the surface of the water. He pointed out that

---

[37]"La grossièreté de la matière verte et le sac dans lequel elle est contenue, ne nous montrent rien d'analogue aux vaisseaux et au sang des animaux, à ceux des plantes et au suc qui coule dedans. Ce sac ressemblerait plutôt à un estomac ou à un intestin et la matière verte à des alimens ou à des excrémens. Mais il est impossible de rien décider encore. Ces grains verts seraient-ils les graines ou les oeufs de ces petits corps? Autre idée qui tombe dans l'esprit. Il faut se préparer à tout dans des cas si extraordinaires. . . . L'on connaît bien des insectes aquatiques et transparens et dont les oeufs qui se voient dans le corps, lui donnent leur couleur. Mais la comparaison n'est pas forte exacte." Ibid., 36–37.

[38]Ibid., 37–38.

[39]Ibid., 43–44.

they shared this characteristic pattern of locomotion with certain aquatic insects and snails. They never seemed to swim or float across the open water, but seemed to need at all times a point of support. For Trembley this implied that there was no similarity between the activity of his enigmatic little bodies "which fix themselves on the side [of the beaker] which has the most light, and that of plants which stretch toward the freshest and warmest air."[40]

Trembley continued to explore the intriguing ability of polyps to move toward the most lighted area of the jar. He asked himself what they would do after he darkened the jar by placing a cardboard box over it. By this experiment he discovered that, even in the dark, his little bodies continued toward the side which had previously been lighted. Trembley did not allow himself to speculate on this curious fact, though it seemed to indicate that polyps instinctively were attracted to light. Not knowing what to make of this capability, he thought that it could perhaps be compared to the ability of birds to migrate.[41]

Réaumur replied to Trembley's letter on 27 February 1741. He still had not seen any live specimens because the first which Trembley had sent him, in an excess of zeal for their preservation in a flask sealed with Spanish wax, arrived dead. In this letter Réaumur stated that he still leaned toward the idea that these little bodies were very sensitive plants.[42]

It was not until March of 1741 that Réaumur could assure Trembley this unusual being was indeed an animal. Trembley at last had been successful in sending live samples to Paris, so that Réaumur could judge this enigmatic discovery with his own eyes. This time rather than taking the excessive precaution of sealing them in a flask, the specimens were carried on horseback in an open container at a pace no faster than a trot. Réaumur, after reporting that he had "observed them by candlelight for an hour and a half with true satisfaction,"

---

[40]Ibid., 47.

[41]Ibid., 48–49. On the subject of Trembley's study of photosensitivity, see Charles W. Bodemer, "Overtures to Behavioral Science: Eighteenth and Nineteenth Century Ideas Relating Light and Animal Behavior," *Epistème*, I (1967), 135–52.

[42]Maurice Trembley, *Correspondance inédite*, 51.

saw no reason to doubt that they were animals and named them for their resemblance to the marine polyp or octopus.[43]

III. THE POLYP'S NATURAL MODES OF REPRODUCTION: BUDDING AND EGGS

Trembley's live polyps reached Paris the same month as the letter in which he announced with obvious delight the climax of his research on the question of how his little bodies naturally reproduced.

On the morning of the 25th of February I saw a little body which was climbing the side of the large glass container in which I keep my supply. In the middle of the body, on the exterior skin, I saw a little excrescence of a dark green color like that of the body when it is contracted.

Trembley isolated this particular polyp from the others and carefully kept track of the changes in the little bud which he suspected would become another organized body.

In the month of July 1740, the time when I only considered these organized bodies occasionally in studying the insects which were in the same glass container, I saw several of these bodies fixed on one another, forming between them different angles. As I believed that they were animals I thought that they fixed themselves on one another as caterpillars do, especially inch-worms. I remembered however that I once suspected that they were really attached and that I thought that this could be a [type of] vegetation. But I did not continue my examination; I was occupied and distracted by other objects. Since the end of August when I started to seriously study these little bodies, I had never seen them on one another. Several times I remembered this winter how I had seen them in the month of July. I confined several in a small space to see if they put themselves on one another. I never saw it. That brought me

---

[43]Ibid., 63.

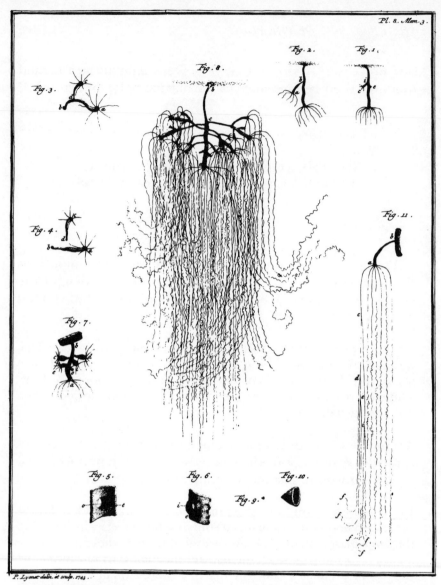

Plate from A. Trembley, *Mémoires*, shows the polyp's natural mode of reproduction by budding. (Leiden edition.)

fig. 1: polyp producing two buds. Fig. 9 shows bud *e* lifesize. Fig. 10 is an enlargement as seen through a microscope.

fig. 2: bud attached to mother

fig. 3–4: bud in the process of detaching itself from the mother

fig. 5–6: two views of the cylinder wall of the mother to which the offspring (also a hollow cylinder) is joined.

fig. 7: polyp with young some time after it has eaten. Buds are swelled with food from the mother.

fig. 8: *Pelmatohydra olgactis*, Trembley's third species, showing two additional generations of offspring

fig. 11: polyp with branched arms

to suspect something and made me conclude so quickly that the excrescence of which I have spoken would become a little body.[44]

Trembley watched this bud grow and produce the characteristic threads at the anterior end. On 15 March, the date of his writing, Trembley stated that he was looking forward to witnessing the complete separation of the new body. At that very moment it was attached by nothing more than a point. Aware that he was witnessing another virgin birth, similar to Bonnet's discovery of parthenogenesis, he vowed that he would not fail to put his newborn in isolation, thus preventing impregnation.[45] Again, Trembley wondered whether this mode of reproduction by budding was not more like plants than animals. "This manner of multiplying must naturally make one think that these organized bodies are plants. If this is the case perhaps they also have a seed."[46]

Réaumur did not enthusiastically accept Trembley's first description of the new phenomenon of budding. Although Trembley's shipment of live polyps enabled him to observe what Trembley described with his own eyes, he questioned whether Trembley's interpretation was correct.

The manner by which these polyps reproduce is yet another marvel, and has no analogy with the manner with which animals perpetuate themselves. What you have observed on this is extremely curious, but one could suspect that you have not observed everything. If the little one, or the egg which the polyp has just produced, attached itself like crayfish attach their eggs under their tail, it could seem united to the body of the mother and when it grew outward, it would only seem that it grew like that of plants. Perhaps the very

---

[44]"Dans le mois de juillet 1740, tems auquel je ne considérais encore les petits corps que par occasion, en étudiant des insectes qui étaient dans le même verre, j'ai vu plusieurs de ces corps qui étaient fixés les uns sur les autres, formant entre eux différens angles. Comme je croyais que c'étaient des animaux, je pensais qu'ils se fixaient les uns sur les autres, comme le font les chenilles, et surtout les arpenteuses. Je me rappelle pourtant que je soupçonnais une fois qu'ils étaient réellement attachés, et que je pensai que ce pouvait être une végétation. Mais je ne continuai point mon examen, j'étais occupé et distrait par d'autres objets. Depuis la fin d'août que me suis mis sérieusement à suivre ces petits corps, je n'en ai jamais vu les uns sur les autres. Je me suis plusieurs fois rappelé cet hiver comment je les avais vus au mois de juillet. J'ai fait en sorte que plusieurs fussent rassemblés dans un petit espace, pour voir si les uns se poseraient sur les autres. Je n'ai jamais pu le voir. Cela m'a porté à soupçonner quelque chose, et c'est ce qui m'a fait si vite penser que l'excrescence dont j'ai parlé deviendrait un petit corps." Ibid., 57.

[45]Ibid., 58.

[46]Ibid.

body of the mother polyp must furnish nourishment to the new-born. These are doubts which are worthy of being investigated, and if they can be, will be by you.[47]

Trembley, referring to the doubts which Réaumur had raised over reproduction by budding, replied that he dared to assert that no egg affixed to the body was involved, but that the young polyp was actually joined to the parent, their two bodies forming a continuous membrane, the green matter within their bodies passing from one to the other. At first, the bud was merely a grain of green matter which appeared to make the skin of the parent slightly raised, the way a button under a stocking causes a slight lump. Trembley was certain that the "excrescence" was interior—not simply attached externally, as Réaumur had suggested. In fact, he wrote, "for the greatest portion of the time of the union of the two polyps they are joined like the branches of plants to the stems from which they emerged."[48] For Trembley, this was an observation of great importance. For this reason he carefully verified it through a variety of experiments and included illustrations which showed the continuous membrane which joined parent and offspring, as well as their common empty central cylinder. The offspring appeared to form out of the material of the parent. First looking like a bulge in the outer skin, it became progressively extended until it resembled its parent. That the parent and child shared a common body, the granules passing without a barrier between them, demonstrated that no egg was involved. Whether Trembley realized the implications that his description of budding might have for the proponents of epigenesis, who argued against a preexisting egg or sperm, cannot be ascertained from a study of the correspondence. It seems likely that the care with which

---

[47]"La manière dont se reproduisent ces polypes est encore un autre prodige, et n'a rien d'analogue à la manière dont les animaux se perpétuent. Ce que vous avez observé sur cela est extrêmement curieux, mais on peut soupçonner que vous n'avez pas tout observé. Si le petit, ou l'oeuf que le polype vient de mettre au jour s'attachait au corps du polype, ou que le polype même l'y attachât comme les écrevisses attachent leurs oeufs sous leur queue, il pourrait paraître uni au corps de la mère, et lorsqu'il croîtrait dessus, il ne semblerait qu'une pousse telle que celle des plantes. Peut-être que c'est le corps même du polype mère qui doit fournir la nourriture au nouveau-né. Ce sont des doutes qui méritent d'être levés et qui, si ils le peuvent être, le seront par vous." Ibid., 65.
[48]Ibid., 72.

he performed his experiments and presented his conclusions stemmed from his desire to be absolutely certain that he was not mistaken.

Nevertheless, even after Trembley had eliminated Réaumur's hypothesis of a concealed egg by a careful description of the structure of the polyp and the process of budding, Réaumur did not give up his search for eggs. In his letters he described to Trembley the research which he and Bernard de Jussieu (1699–1777) were carrying out on the plumed, or tufted, polyp, actually in modern terms not a hydra, but the bryozoan, the *Lophopus*. Their observations suggested that polyps produce eggs which are analogous to seeds in plants. In his letter of 30 August 1741 Réaumur referred to the black grains which he had observed. He questioned whether these grains were really excrement.

I have already reviewed all that you have written to me, but I am uncertain if the black grains which you have called excrement are the same black grains which I have seen—generally three or four in the body of each polyp. I am very troubled to find out what these grains are which I cannot regard as the remains of digested matter.[49]

What baffled Réaumur was that these grains at times appeared to be mobile like little fish swimming in the body of the polyp. Could they be little parasitic animals, or even perhaps the eggs or young of the polyp?[50]

In his letter of 9 November Trembley answered Réaumur's queries about the black grains. "The black grains which you have seen have certainly no relation with those I judged to be digested matter. There is never but a single brown grain of the sort of which I speak."[51] Moreover, this brown grain was immobile, subject only to the mechanical movement of the body. He added that what Réaumur had observed had definitely excited his curiosity. "I have not seen similar ones, though I have looked with care since the reception of your letter, but a long time before I received it, I did notice some white ones to which I have already given a great deal of

---

[49]Ibid., 104.
[50]Ibid., 105.
[51]Ibid., 107.

attention."[52] Trembley noted that these bodies were perfect spheres and had a transparent skin. Sometimes they appeared to be filled with white matter. They were in continual movement, and each seemed to be able to move independently of the others. Trembley was even able to observe one pass from the body of one into another through their common base. "I do not know what to say these grains are. It occurred to me that they were perhaps animals which live in the polyp or even the eggs of the polyp. The latter idea appears to me somewhat likely and, if it is substantiated, [we may conclude that] the plumed polyps multiply both by *eggs* and by *shoots*."[53] However, Trembley was far from convinced of this idea. He had not found them in all polyps—not even in a great number of them.

Réaumur did not have the same reservations about the presence of eggs in this particular species of polyp. He discussed the polyp in detail in his *Histoire des insectes* before Trembley had completed his work for publication. Réaumur reasoned that it was not surprising that polyps, besides reproducing by budding like plants, should also produce eggs analogous to seeds in plants. He recounted how with Bernard de Jussieu he had indeed seen brown grains in the plumed polyp produced shortly before the polyp died.[54]

Oddly, even after Réaumur's declaration in print that polyps lay eggs, Trembley did not fully subscribe to the idea. On 6 June 1743 he voiced only the suspicion that polyps could come from eggs,[55] while Réaumur in his letter of 26 July declared unequivocally that he had seen polyps hatch from eggs. He was sure that the polyps discovered in a beaker of water after his entire supply had presumably died, had hatched from eggs laid just before they perished.[56] Though on 8 August Trembley thanked Réaumur for these curious facts, he still did not think that he had conclusive evidence. "I have never seen one of them come out of the egg."[57]

On 17 December 1743 Réaumur informed Trembley that

---

[52]Ibid., 111.
[53]Ibid., 112.
[54]Réaumur, *Histoire des insectes* 6: lxxvi.
[55]Maurice Trembley, *Correspondance inédite*, 168.
[56]Ibid., 170.
[57]Ibid., 174.

he would be interested in the observations which Bernard de Jussieu and his companions had made on another species, the "horned" polyp.

They found in a quantity of these polyps a little vessel adherent to their body and which was filled with eggs. It would be happy for the analogy, if the horned polyp like the plumed polyp were oviparous and viviparous. I have difficulty in believing that these horned polyps, if they have eggs, could hide them from you.[58]

Réaumur added that de Jussieu's observations were not conclusive since he did not actually see the eggs hatch. In his letter of 23 January 1774 Trembley referred to de Jussieu's eggs. Observations made by the famous botanist of the Jardin des Plantes were not to be taken lightly. He said that he had also observed the vessels which appeared to contain eggs, but it had not been possible for him to verify whether they were indeed eggs, and he still had doubts about it.[59]

As late as 1744, Trembley was still not fully convinced that polyps produced eggs. Though the question was raised and a description of the vessels (ovaries, in fact) which appeared to contain eggs was given, Trembley was reluctant to draw this conclusion because he had not actually observed a polyp hatch from one of these spherical bodies.[60] However, in a now-missing letter written in the fall of 1745 to Réaumur, Trembley again returned to the problem of eggs and apparently sent samples of brown spherical bodies of the plumed polyp which he wished Réaumur to compare to the ones which he had observed with Bernard de Jussieu. Réaumur confirmed that they were the same and that undoubtedly Trembley would see little polyps hatch from them in the spring.[61] In the last letter to Réaumur which has been found, Trembley at last accepted the presence of eggs. The long letter of December 1746 examined yet more species of polyps and carefully described their modes of reproduction. There was no doubt left in Trembley's mind that polyps reproduce both by budding and by eggs.[62]

---

[58]Ibid., 181.
[59]Ibid., 184.
[60]See John R. Baker, *Abraham Trembley of Geneva*, 90–91.
[61]Maurice Trembley, *Correspondance inédite*, 249.
[62]Ibid., 275.

Though Réaumur was correct in maintaining that polyps produce eggs, he made a mistake in his description of the plumed polyp *(Lophopus)*, a mistake that reveals his fundamental misunderstanding of the structure of the polyp which Trembley was in the process of elucidating. Réaumur thought that the plumed polyp produced tubes, similar to nests or cells, piled one upon the other and within which the polyps resided.[63] Réaumur made the same mistake in his description of corals.[64] He believed that the animals lived within structures analogous to bee hives. Thus, for Réaumur, polyps like many other animals which he had studied, demonstrated a certain "genius" for construction.

This was a fundamental misunderstanding of the plumed polyp's *(Lophopus)* structure.[65] Trembley correctly revealed that what appeared to be lodgings, built by the polyps, were in fact part of the living bodies of the polyps. Even more important, Trembley realized that the living relationship between the mother and its young was one of interdependence. They shared a common body trunk and were connected with each other by a continuous living membrane. Trembley carefully demonstrated that the foodstuffs taken in by the mouth of either mother or child could pass from the body of one to the other.[66] The colonial structure of the plumed polyp *(Lophopus)* which Réaumur had envisioned as a series of lodgings piled one on the other, was actually a living mass. Trembley pointed out that the living membrane was never broken, since the child did not separate from the mother as it did in the case of other polyps, but remained attached. Both mother and child were united in structure and function throughout

[63]Réaumur, *Histoire des insectes* 6: lxx. See also John R. Baker, *Abraham Trembley of Geneva*, 125.

[64]As a consequence of Trembley's discovery of the polyp, Réaumur realized that what Luigi Marsigli had described as flowers of sea corals were in fact little animals. Jean-André Peyssonnel had argued this view in a paper sent to the Academy of Sciences in 1727. Since Réaumur supported Marsigli's opinion, he did not have Peyssonnel's paper published, though he had informed the Academy of its contents, withholding the name of its author to prevent him embarrassment for advancing an idea which appeared to Réaumur so misguided. In the introduction to Volume Six of his *Histoire des insectes*, lxxii–lxxix, Réaumur vindicated himself by giving Peyssonel full credit.

[65]Trembley's letter of 9 November 1741 is devoted to a full discussion of the structure of the plumed or tufted polyp. Maurice Trembley, *Correspondance inédite*, 107–14.

[66]Ibid., 111. See also sketch, 112.

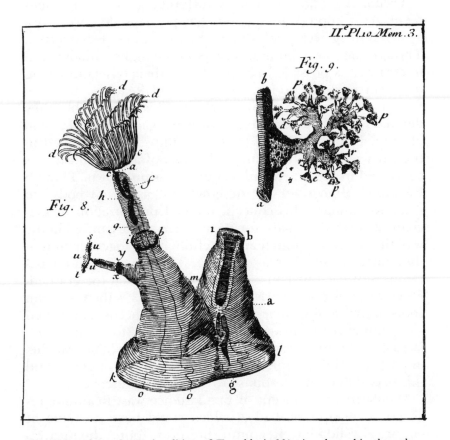

Illustrations from the Paris edition of Trembley's *Mémoires* shows his plumed or tufted polyp, the bryozoan *Lophopus*. Note the animal's colonial structure in the lower illustration which led Trembley to compare it to coral, its salt water counterpart, heretofore believed to be a plant. Réaumur thought that these polyps produced *polypiers* within which they resided. Trembley established that rather than an external lodging, the outer membrane was part of the living body of the polyp.

(Permission to reproduce illustrations from the Paris edition, courtesy of the Department of Special Collections, Case Western Reserve University Libraries.)

their lives, and thus created the branched appearance of the colony.

Trembley's and Réaumur's differences on these issues seem great enough to suggest that perhaps they were not approaching the problem from the same perspective. Though they were both observers of the highest quality and integrity, it is possible that they based their observations on different preconceptions. Trembley and Réaumur both referred to the polyp as a little machine. This, presumably, they derived from their common Cartesian legacy. The discussion of animals in mechanical terms was universally accepted in the early eighteenth century. The idea of an "organism," as conceived in the nineteenth century, was unknown. Though Réaumur often used the designation "organized body" to refer to animals, this concept implied organization in mechanical terms. Animals, for Réaumur, were organized in the way the parts of a watch are carefully fitted together so that they operate together as a whole. The integrity and separateness of each machine of nature were taken for granted. However, it is not surprising that by the 1730s the Cartesian roots of the "animal machine" concept had become entwined with influences from other philosophical systems and intellectual traditions. Réaumur was influenced by Malebranche, in particular by his concept of the preexistence of germs and this bias shaped his response to Trembley's discovery.

Malebranche's concept of preexistence was, Réaumur realized, a metaphysical idea, which could have no hope of empirical validation. Réaumur, however, found the idea rationally acceptable: "The tree which is reduced to such a small size in this seed, could have been of an even smaller size; there could have been a time when it was enclosed in a seed of an imperceptible size—a time when it was as small in comparison to an ordinary seed as this seed is small with respect to the tallest elm. Only the imagination is alarmed by this; reason, as soon as it is convinced of the divisibility of matter to infinity, is not at all shocked by these incredibly small sizes."[67] Réaumur's insistence that the polyp must produce

---

[67]"L'arbre qui est réduit si en petit dans cette graine, a pû être d'une petitesse prodigieusement plus grande; il peut y avoir eu des temps où il étoit renfermé dans une graine d'une grosseur insensible, des temps où il étoit aussi petit par rapport

eggs is related to his belief in preexistence. For Réaumur, the eggs of animals were analogous to the seeds of plants. Such infinite diminution was necessary for the perpetuation of animals through the many centuries after Creation. In the second volume of his *Histoire des insectes* Réaumur attempted to give an empirical justification for his belief. He discussed his success in varnishing insect eggs, and then removing the varnish after several years and watching living insects emerge. He reasoned that if both seeds and eggs could be artificially preserved, and then allowed to germinate or develop by providing the appropriate conditions, it was conceivable that nature operated in the same way, but on a much larger time scale. The life span of an insect appeared short only until it was realized that it could have lived several years in an egg without growing and that this period could be extended indefinitely: "we can conceive that there was a time when this insect, prodigiously smaller than it is in the egg, was enclosed under an envelope of an indeterminable smallness where it lived without expanding and without growing, and it could be there centuries or many centuries without perceptible growth."[68]

For both Réaumur and Malebranche it was only the egg or seed which ensured that the preformed individual was carried in a sort of suspended animation until the parts began to expand under favorable conditions. This was a mechanical idea, since the life of an individual was conceived to be a series of steps of growth and shrinking that could be speeded up or slowed, but could not be otherwise changed. The pattern was fixed; only external conditions could vary. Gener-

à ce qu'il est dans une graine ordinaire, qu'il est petit dans cette graine par rapport au plus grand orme. L'imagination seul s'effraye ici, la raison n'est point étonnée de toutes ces enormes petitesses, dès qu'elle s'est convaincuë de la divisibilité de la matière à l'infini." Réaumur, *Histoire des insectes* 2: 31. I am indebted to Jean Torlais for showing that Réaumur was influenced by Malebranche. See "Réaumur philosophe" in *La vie et l'oeuvre de Réaumur* (Paris: Presses Universitaires de France, 1962), 145–65.

[68]"Dès que nous voyons qu'un insecte qui ne vit pour nous que quelques mois, peut avoir vécu auparavant plusieurs années dans un oeuf, parce qu'il n'y croissoit point, nous pouvons concevoir qu'il y a eu des temps où cet insecte prodigieusement plus petit qu'il ne l'est dans l'oeuf, étoit renfermé sous une enveloppe d'une petitesse indéterminable, où il vivoit sans s'étendre & sans se développer, & qu'il y a pû être renfermé pendant des siècles & des suites de siècles sans croître sensiblement. Les plantes sont propres à nous disposer à nous révolter moins contre une idée qui a quelque chose d'effrayant." Réaumur, *Histoire des insectes* 2: 30.

ation was not thought to be creative in the sense that something new was formed. It was only that the expansion in size made development visible.

Trembley's denial that any egg was involved in the phenomenon of budding was a blatant contradiction of Réaumur's first "certain principle" of natural history that there is "no species of insect which is neither oviparous nor viviparous, that does not perpetuate itself either by laying eggs, or by bringing forth living young."[69] Neither reproduction by cuttings nor the phenomenon of budding adhered to this rule. Though budding indicated that the polyp might be viviparous, even viviparous animals were thought to come from eggs that developed internally.[70] This conviction is why Réaumur was so incredulous when Trembley's observations established that in budding the "excrescence" that developed into the bud was a part of the skin of the mother rather than being something externally attached, in the way to be expected of an egg. Rather, the bodies of mother and child were connected, as fingers on a glove, by a continuous membrane.

Moreover, the regeneration of whole individuals from cuttings must have seemed to Réaumur to come dangerously close to reintroducing the old idea of spontaneous generation. Réaumur regarded a belief in spontaneous generation as one of the greatest obstacles to the serious study of the natural history of insects, "because as soon as it is believed that they come from rotting matter, the most interesting part of their history, all that relates to the subject of how they perpetuate themselves, does not seem to need to be studied."[71] In fact, Réaumur's explicit statement of his principles of natural history in the introduction to Volume Two of the *Histoire des insectes*[72] was a consequence of his clash with the Jesuit fathers of the *Journal de Trévoux* over his statement in the introduction to Volume One of the *Histoire des insectes*

---

[69]"Qu'il n'y a aucune espèce d'insecte qui ne soit ovipare ou vivipare; qui ne se perpétue, soit en pondant des oeufs, soit en mettant au jour des petits vivans." Ibid., xxxix.

[70]The discovery of the mammalian ovum, claimed by Regnier de Graaf in 1672, strengthened the belief that all life comes from eggs. In fact, he had not discovered the ovum, but the follicles of the ovary from which the eggs had emerged.

[71]Réaumur, *Histoire des insectes* 2: xvi.

[72]Ibid., xxxix–xli.

(1734) that Athanasius Kircher was among those who sub-
scribed to the false idea of the ancients that generation could
spontaneously occur from rotting vegetable matter. Réaumur's Jesuit critics defended Kircher's idea that pulverized
worms could spontaneously regenerate. Réaumur carefully
repeated Kircher's experiments and proved that Kircher's
recipe for the regeneration of cut-up earthworms was impossible.[73]

Yet here was Trembley, only a few years after Réaumur
had refuted Kircher's "system," reintroducing Kircher's very
idea that cut-up "worms" could regenerate. The difference
was that Trembley had unassailable experimental evidence
to back up the argument for regeneration. But regeneration
upset the orderly pattern presupposed by a belief in preexistence. Regeneration and budding seemed to suggest that
matter might have the power to organize itself, without the
interposition of an egg. This contradicted Réaumur's most
cherished belief that chance played no part in the generation
of animals. "If one does not stop at gross appearances," Réaumur had written in his earlier volume, "if one is attentive to
all the circumstances on which the birth of insects depends,
one will be convinced that chance has no more role in it any
more than it has in that of large animals or even that of
man."[74]

Trembley's discovery of regeneration in the polyp undermined the whole idea of the animal machine. The functional
completeness dictated by the analogy with a watch mandated
that if the principal organs were severed from each other,
death would result. Obviously no watch cut in half could
continue to keep time. Like Réaumur, Trembley believed in
the necessary functional unity of the animal. He had expected
his little being to die after cutting it, but when the severed
parts regenerated, he systematically set about to examine this
phenomenon, trying to put aside his preconceptions. Since
the reproduction of aphids had upset the general rules, perhaps the polyp was another exception. Could a clear line

[73]See Réaumur's lengthy discussion, ibid., xvi ff.

[74]"Si on ne s'arrête pas à des apparences grossieres, si on est attentif à toutes les
circonstances d'où dépend la naissance des insectes, on sera convaincu que l'hazard
n'y a pas plus de part qu'il n'en a à celle des grandes animaux, à celle de l'homme
même." Ibid., xxiv.

dividing the animal and plant realms be made? Had he discovered a new class of organized body *between* the two kingdoms? The spontaneity of the polyp's behavior seemed to suggest a soul. Would this stir up the debate over animal soul?[75]

IV. BOERHAAVE'S DEFINITION AND TURNING THE POLYP INSIDE OUT

Seeking a way to settle his questions, Trembley recalled a definition of the difference between plants and animals found in Boerhaave's *Elementa chemiae* (1732), a work with which Trembley appears to have been thoroughly familiar. In an early letter to Réaumur he wrote:

It is currently rather difficult to mark the difference between plants and animals. Mr. Boerhave indicates principally the distinction between the locations of the parts through which they draw their nourishment. According to him, plants have external roots and animals, interior ones. It is not possible to say yet how our little bodies nourish themselves. I have some possibilities which lead me to believe that these organized bodies are a particular species of a new class of organized body until now unknown and intermediate between animals and plants.[76]

Trembley also referred to Boerhaave's definition in his *Mémoires.*

---

[75]"Les découvertes que l'on a faites sur les plantes et les animaux nous ont appris qu'il y a entre eux de très grands rapports. Peut-être qu'à force de les étudier on en trouvera tant que cette grande distinction qu'on met entr'eux s'évanouira. L'on ne tardera peut-être pas à agiter la question si les plantes ont une âme." M. Trembley, *Correspondance inédite,* 59.

[76]"Il est actuellement assez difficile de marquer la différence qui est entre les plantes et les animaux. Monsieur Boerhave indique principalement celle qui est entre la situation des parties par où ils tirent leur nourriture. Les plantes, suivant lui, ont des racines extérieures, et les animaux des racines intérieures. Il ne m'est pas possible de dire encore comment nos petits corps se nourissent. J'ai quelques probabilités qui laisseraient croire que c'est comme les animaux. Je suis quelquefois tenté de croire que c'est comme les animaux. Je suis quelquefois tenté de croire que ces corps organisés sont une espèce particulière d'une nouvelle classe de corps organisés, jusqu'ici inconnue, et mitoïenne entre les animaux et les plantes." Ibid., 60–61. Boerhaave's definition is discussed by François Delaporte, *Nature's Second Kingdom,* trans. A. Goldhammer (Cambridge, Mass.: MIT Press, 1981), 80.

One can hardly cite on this occasion any more respectable authority than that of the celebrated Boerhave. How this great man labored to study Plants and Animals! However it appears that he has found only a single general and essential difference between these two classes of organized bodies: this is what can be seen at the beginning of his Chemistry in the article where he considers plants and animals. This difference consists in the manner with which plants and animals draw their nourishment. *The nourishment of plants*, says Mr. Boerhave, *is drawn by exterior roots, and that of animals by interior roots. This exterior part, called the root, which draws food from the body in which it is situated is sufficient to distinguish a Plant from any Animal known up to now.* And in the definition which he gives of the body of an animal, he says, *that it has within it vessels in the guise of roots by which it draws the food material.* Finally, after having insisted on the comparison between a Plant and an Animal, he adds: *in this consists the relation and the difference which there is between a plant and an animal.*[77]

In his chemistry Boerhaave drew on the iatro-mechanistic tradition of Santorio Sanctorius and Lorenzo Bellini to describe the life functions of animals in terms of the mechanical movement of fluids. He conceived of an animal as a kind of hydraulic machine "which subsists by a constant and determin'd motion of humours through its vessels, and which contains within itself certain vessels, like the roots of Vegetables, by which it draws in that nutriment, which supports its being, and increases its magnitude."[78] He thought that the bodies of animals were made up of a hierarchy of vessels which

---

[77]"On ne peut guères citer à cette occasion d'Autorité plus respectable que celle du célèbre Boerhave. Combien ce grand homme n'avoit-il pas travaillé à étudier les Plantes, & des Animaux? Cependant, il paroît qu'il n'a trouvé qu'une seule différence générale & essentielle entre ces deux Classes de corps organisés. C'est ce qu'on peut voir au commencement de sa *Chymie*, dans les articles où il traite des Plantes & des Animaux. Cette différence consiste dans la maniére dont les Plantes & les Animaux tirent leurs alimens. *Les alimens des Plantes*, dit Mr. Boerhave, *sont tirés par des racines extérieures, & ceux des Animaux par des racines intérieures. Cette partie extérieure, appellée racine, qui tire les alimens du corps dans lequel elle est placée, distingue assés une Plante de tout Animal connu jusqu'à présent.* Et, dans la définition qu'il donne du corps d'un Animal, il dit, *qu'il a en dedans des vaisseaux, en guise de racines, par lesquels il tire la matiére des alimens.* Ensuite, après avoir encore insisté sur la comparaison d'une Plante & d'un Animal, il ajoute: *c'est en cela, que consiste le rapport & la différence qu'il y a entre un Végétable & un Animal.*" Abraham Trembley, *Mémoires*, 306–307. Trembley cited the Leiden edition of 1732, 57–64. This was the authorized version of Boerhaave's chemistry, prepared under his supervision, after several editions of students' notes had been published. See F.W. Gibbs, "Boerhaave's Chemical Writings," *Ambix* 6(1958): 117–35 for information about the various editions of Boerhaave's works.

[78]Boerhaave, *Elements of Chemistry*, tr. Timothy Dallowe (London: J. & J. Pemberton, 1735), 40.

carried the various fluids of the body: the sanguiniferous
vessels for the venous and arterial fluids, the lymphatics, and
finally the lacteal and mesenteric vessels through which the
products of digestion moved from the gut eventually into
the blood. Just as in plants which drew their nourishment
from the earth through roots, the lacteal vessels were the
medium through which the chyle in the stomach was strained
to become eventually blended into the blood, a process which
he thought took place in the lungs. That the lacteal vessels
might be so small that they could not always be observed did
not prevent Boerhaave from postulating that they must exist
as the locus for the transformations of the various particles
of which the body was composed.[79] Trembley's use of Boerhaave's definition as the criterion by which to judge whether
his discovery was animal or plant is important because it
shows that Boerhaave's influence in this respect far outweighed the authority of Réaumur.

Réaumur appears to have concluded that the polyp was an
animal from its locomotion. However, Trembley's familiarity
with Boerhaave's criterion based on nutrition made him determined to study the structure of the polyp more carefully.
He still did not know how his little beings took food! Two
months after Réaumur had declared that the polyp was an
animal, Trembley could at last answer the question of how
the polyp sustained itself. He had watched one ensnare a tiny
eel with its arms and stuff it into the central cavity which he
now realized was its stomach. He could then write to Réaumur with satisfaction: "They are carnivorous, and certainly
very avid."[80] He was at last convinced that the creature was
indeed an animal.

One of the reasons for Trembley's delay in publication of
his discovery, despite Réaumur's urging, appears to have
been his determination to understand the polyp's structure.[81]
It was only gradually through continuous observation that

---

[79]See Lester S. King, *The Growth of Medical Thought* (Chicago: University of Chicago
Press, 1963), 181. Information on the lymphatic system was difficult to find. I am
indebted to Elaine Robson for directing me to an article by Nellie Eales, "The History
of the Lymphatic System, with Special Reference to the Hunter-Monro Controversy," *Journal of the History of Medicine and the Allied Sciences* 29(1974): 280–94.
[80]M. Trembley, *Correspondance inédite*, 78.
[81]Ibid., 87.

Trembley was able to identify the various parts of the polyp and their functions. For example, at first Trembley had thought that the tentacles were possibly roots through which it drew its nourishment. Then he considered whether the threads were perhaps legs, since they obviously assisted it in walking. It was not until Trembley correctly ascertained that the tentacles were arms which assisted the animal in seizing its prey that he was sure that he had answered the question of how it took food. He could then refer to the central cavity as the stomach.

The reason why he did not realize that the tentacles ringed the hydra's mouth has been explained with great insight by Richard Campbell in his scientific paper, "Does a Hydra have a Mouth (when it is closed)?"[82] He explains that the hydra can go many months without eating. When the hydra's mouth is closed, tissue completely covers the opening, making a mouth very difficult to imagine and impossible to see. It was only when Trembley caught the polyp in the act of ensnaring an eel with its tentacles that he was able to see what had looked like solid tissue open up.

How was the polyp able to absorb nutriments, if there were no structures within which food was prepared for absorption? Rather than lacteal vessels, Trembley observed that the body appeared to be composed of tiny granules. He found these granules on the internal and external surfaces of the polyp, as well as spread throughout its entire thickness. These granules were the *only* structures which he was able to find within the body, and he felt justified in giving them close attention because their study furnished him with "the only ideas I have about how Polyps are organized."[83] He studied the granules carefully, but found no vessels within which they were held.

The issue of how digestion occurred, and his search for lacteal vessels in this perplexingly simple creature drove him to an ever more careful series of experiments. The first was to cut the polyp, not transversely as he had done first to make a cross-section, but down the body, splitting the double mem-

---

[82]See Richard Campbell, "Does a Hydra have a Mouth (when it is closed)?" *From Trembley's Polyps to New Directions in Research on Hydra* in *Archives des Sciences* (Genève, 1985) 38: 359–69.
[83]A. Trembley, *Mémoires*, 60.

brane the length of the stomach. This left two equal halves, each with half the tentacles and mouth, but deprived the animal of its digestive cavity. Thus, he wrote to Réaumur, "I arrived at a fact which surprised me more than all that I have seen up to the present in the polyps."[84] In less than a day, each half had reformed into a polyp, complete with central cavity and by afternoon, Trembley witnessed each feed on a small eel.

Why Trembley should have regarded the lengthwise splitting of the polyp with greater astonishment than that of transverse sectioning is perplexing. It appears that he believed that this lengthwise sectioning should destroy the polyp's "animal oeconomy," to use eighteenth-century terminology. That the regrowth of the cross sections should have seemed less disruptive of the animal's structure must be related to his belief that the ability to digest food was of prime importance in the functioning of the animal machine. Without that "constant and determin'd motion of humours through its vessels," the hydraulics of the animal machine should have been irreparably upset, yet this did not happen at all. The polyp reformed its digestive cavity as easily as it had reformed head and tail in Trembley's earlier experiments. Had he known at the time that even the smallest section could regenerate, the experiment would have seemed less surprising. Trembley began with the assumption that he would be able to determine experimentally at what point sectioning would kill the animal. He discovered that he could go so far as to chop it into little pieces and that it would always reform into a being with a central digestive cavity.[85]

The perplexing phenomena surrounding the question of nutrition and his exploration of structure, led Trembley to his most famous experiment of all—that of *retournement,* or turning the polyp inside out. This experiment required such dexterity that researchers in the early nineteenth century could not repeat it and doubted that Trembley had actually performed it. More recent scientific opinion adds new questions, casts doubt on Trembley's interpretation of his experiment, but as yet no consensus has been reached on exactly

---

[84]M. Trembley, *Correspondance inédite,* 87.
[85]A. Trembley, *Mémoires,* 248.

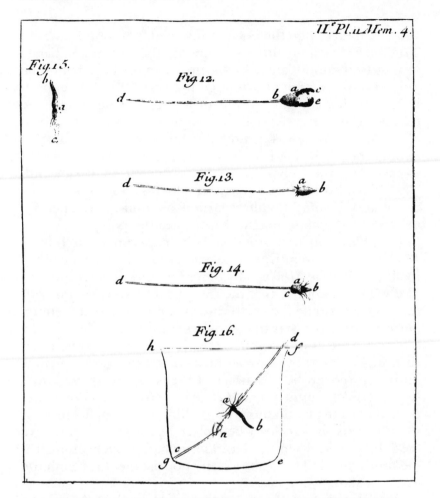

Trembley's illustration of turning the polyp inside out. (Paris edition.)
Fig. 12: A boar's bristle is pushed against the posterior end of the polyp which has just been fed the larva of a crane-fly. This forces the mouth to open wide to enable the posterior end to be pushed through the mouth.
Fig. 13: The bristle is now enveloped by the polyp, whose exterior skin is now on the inside.
Fig. 14: Here the mouth portion of the polyp has not been completely inverted.
Fig. 15 (upper left): The polyp as it appears completely inverted.
Fig. 16: The inverted polyp spitted near its lips on a bristle. A knot, n, has been placed below the polyp so that it is prevented from slipping down on the bristle to the bottom of the jar. So threaded, Trembley claimed that the polyp could continue to carry on its life functions.
(Courtesy of Department of Special Collections, Case Western Reserve University Libraries.)

what happens when the polyp is inverted.[86] So important did Trembley consider this experiment, that he immediately sought to demonstrate it in the presence of those whom he judged most competent to judge the veracity of the experiment and his conclusions "in order to be able to cite the testimony of other witnesses besides myself to prove the truth of a Fact as strange as this one."[87] Therefore, in addition to Lyonet, he chose his friend Allamand, who also succeeded in inverting the polyp and then turning it right side out. Trembley also called on Bernard Albinus, Boerhaave's successor and the most highly regarded anatomist of the period, to confirm the experiment of inverting the polyp.

Trembley had first attempted the experiment in July of 1741, that is the month after he had discovered how the polyp took food. In the *Mémoires* he described how he first thought of it when he noticed that the small granules in the lining of the stomach swelled up with nutritive juices after a feeding. Since he thought that the granules on the exterior and the interior surfaces of the skin were identical, he reasoned that each surface ought to have a similar ability to absorb nutritive juices. Moreover, he was aware of the (soon to be discarded) hypothesis of Count Luigi Marsigli in *Histoire de la mer* that certain marine plants absorb nourishment externally through small glands or vesicles when placed in a sort of nutritive bath. However, when he placed his polyps in such a nutritive medium, they failed to absorb any nourishment through the external skin.[88]

Since this approach was not successful, he then thought of inverting the polyp so that the external skin would be on the

---

[86] In fact, the inner and outer walls of the hydra's body (endoderm and ectoderm) are not strictly interchangeable. At present considerable differences of opinion exist on the exact interpretation of Trembley's experiment. See the author's article, "Trembley's Experiment of Turning the Polyp Inside Out and the Influence of Dutch Science," in *From Trembley's Polyps*, 320–34. John Baker mistakenly thought that this experiment, which he called "reversal," could be explained by the migration of the cells between the two cell layers. However, more recent opinion disputes this. See Robert L. Roudabush, "Phenomenon of Regeneration in Everted Hydra," *The Biological Bulletin* 64 (1933): 253–58; and Martin Macklin, "Reversal of Cell Layers in Hydra: A Critical Re-appraisal," *The Biological Bulletin* 134 (1968): 465–72. I wish to thank Howard Lenhoff for assistance in understanding the scientific aspects of this experiment and for calling my attention to important scientific references.

[87] A. Trembley, *Mémoires*, 264.

[88] Ibid., 254. Trembley modestly refrained from telling his readers that it was his own discovery of the polyp that had led to a reassessment of Marsigli's "system."

inside. "I thought of turning Polyps inside out after noticing a Fact detailed in the second Memoir: namely, that the granules or vesicles which adorn the entire skin of these animals fill up with nutritive juices. The thought came to me that if the granules which were on the external surface of the skin were closer to this nutrient juice, they would be the first to become filled with it, and that perhaps the Polyp would be nourished as thoroughly as when the juice passes first into the vesicles lining the walls of the stomach."[89] However, his first attempts to turn the polyp inside out failed. It was not until he thought to try it after the polyp had been well fed with a small worm that he succeeded. The process of inversion was similar to turning the finger of a glove inside out.

To invert the polyp, Trembley first fed it an eel. The bulk of the eel caused the mouth and stomach to become somewhat stretched. Then, carefully cradling a polyp in the palm of his hand in a drop of water, he pushed on the posterior end with the bristles of a brush. As the end was pushed toward the mouth, the eel was wedged against it, forcing it open. The second step was to remove the polyp to the moistened edge of his hand, while continuing to force the body to contract and the mouth to open wider. In step three, Trembley took a thick boar's bristle with a blunt end and, holding it "as one does a lancet to let blood," he pushed the posterior end so that it entered the stomach and emerged through the mouth. "One can visualize that the inverted Polyp then envelops the end of the boar's bristle which is now lodged within it; the exterior surface of the Polyp has become the interior surface touching the bristle, and the interior surface has become the exterior."[90] Finally, to prevent the polyp from turning itself back, Trembley pierced its body near its lips with

[89]"J'ai pensé à retourner des Polypes, après avoir remarqué un Fait, qui se trouve detaillé dans le second Mémoire: savoir, que les grains, ou vessicules, dont toute la peau de ces Animaux est garnie, se remplissent de suc nourricier. Il me vint dans l'esprit, que, si les vessicules, qui étoient à la superficie extérieure de la peau, se trouvoient les plus près de ce suc nourricier, elles s'en rempliroient les premiéres, & que le Polype se nourriroit, peut-être, aussi bien que lorsque le suc nourricier passe d'abord dans les vessicules qui tapissent les parois de l'estomac." Ibid.

[90]"On conçoit qu'il couvre alors le bout de soie de sanglier, qui est logé dans le Polype retourné; que la superficie extérieure du Polype est devenue intérieure; que cette superficie touche celle de la soie de sanglier, & que l'intérieure est devenue extérieure." Ibid., 257.

a thin boar's bristle which he tied at one end. He then situated the spitted polyp in one of his powder jars in such a manner that it was prevented from touching the sides or the bottom of the jar and thus could not get the leverage to turn itself back. Thus suspended on the bristle, the polyp continued to live, eat and reproduce.

Trembley's careful study of the polyp did not reveal the analogous structures to be expected in the body of an animal. Trembley found that polyps had no nutritive vessels at all, neither roots nor the digestive vessels that Boerhaave had claimed must be found in animals. Trembley did not rule out the possibility that such structures might be so small as to be imperceptible, but he rejected Boerhaave's assertion that such structures *must* exist. He concluded that the polyp was itself a single nutritive vessel consisting of a single skin, "shaped in the form of a tube, or gut, open at both its ends."[91] In contrast to other animals, like caterpillars or different species of worms, which have various vessels within them, polyps had only one whose surface comprised the very surface of the animal. Nevertheless, in denying that polyps had a structure comparable to other so-called insects, Trembley still maintained that polyps were possessed of a structure and it was this structure that was the key to defining the polyp as an animal.

Through the experiment of turning the polyp inside out, Trembley concluded that, though there were no distinguishable smaller vessels, the fact that the polyp drew its nourishment from within like other more complex animals indicated according to Boerhaave's definition that the polyp was a simple animal. Thus, he rejected the view that the polyp ought to be considered an intermediate being, *between* the animal and vegetable realms. While future research might determine the creation of new classes, the present state of knowledge indicated that the polyp was better classified as an animal.

---

[91]"Tout l'animal ne consiste que dans une seule peau, disposée en forme de tuiau, ou de boiau, ouvert par ses deux extrémités. Lorsqu'on ouvre d'autres Animaux, des Chenilles, diverses espéces de Vers, on leur trouve différens vaisseaux; mais en ouvrant les Polypes, on n'en trouve absolument qu'un, aussi long que le Polype, ou plutôt, comme je l'ai déja dit, tout cet Animal ne paroit former qu'un vaisseau, dont la superficie extérieure est la superficie même de l'animal." Ibid., 52.

In order to be able to decide that such an organized body is neither plant nor animal, but ought to be placed in an intermediate class [classe mitoïenne] between animals and plants, in order to be able to decide, I say, it would be necessary to know precisely all the properties to which animals & plants are susceptible. As we have seen, we are very far from this knowledge. It will only be when we have arrived at this that we will be able to make other classes of organized bodies. In waiting it is much more natural to consider Polyps & various other organized bodies which have received the name of Zoophytes, as animals which have more remarkable similarities with plants than do others.[92]

Trembley's exploration of the question of whether the polyp was animal or plant and his willingness to entertain the possibility that intermediate classes might eventually be discovered, points to the possible influence of the concept of the great chain of being. The chain of being was discussed by a variety of British authors in the early eighteenth century, particularly Pope, Shaftsbury, Addison, and Locke. Indeed, although there is nothing in Trembley's early letters or his Mémoires to tie him directly to specific passages where these authors discuss the chain of being, so common was the idea that it would be surprising if Trembley was not to an extent influenced by it, especially since he was fluent in English and had read extensively in British natural theology. That, in the end, he rejected the view that the polyp provided the link between the animal and plant realms does not deny the initial impetus that it gave to his careful study of its elusive structure.

Fully aware that philosophical notions could obscure the clear view of nature, nevertheless Trembley was attracted to the writings of John Locke. His contemplation of one of the most provocative of Locke's philosophical questions, whether matter might have the capacity to think, reveals that Trem-

---

[92]"Afin de pouvoir décider, que tel corps organisé, n'est ni plante, ni Animal, mais doit être placé dans une classe mitoïenne entre les Animaux & les Plantes; afin, dis-je, de le pouvoir décider, il faudroit connoitre précisément toutes les propriétés dont les Plantes & les Animaux sont susceptibles. Nous avons vu ci-dessus, que nous étions fort éloignés de cette connoissance. Ce n'est que lorsqu'on y sera parvenu, qu'on pourra faire d'autres classes de corps organisés. En attendant, il est beaucoup plus naturel de regarder les Polypes, & divers autres corps organisés qui ont reçu le nom de Zoophites, comme des Animaux qui ont plus de rapports remarquables avec les Plantes, que d'autres." Ibid., 308.

bley was not merely a shallow empiricist. Under the heading "Soul," he wrote:

Look at what Mr. Locke thinks about Thinking; he believes that it is a property which God could have given to Matter. Locke, Book IV, chap. 3 & 6. . . . If God could give to matter the power of thinking, what we call the soul and which is conceived of clearly as indivisible, could nevertheless be divided. Therefore, since there is no part of matter which cannot be divided, I could conceive that the parts of matter which are in me, which I call my soul and which can think, could be divided, and I could then conceive that my soul can be subdivided. That seems to me a little absurd.[93]

Although Trembley found Locke's conclusion absurd, because that would imply that the human soul was divisible, it is significant that he was aware of the philosophical issues concerning the properties of matter that perplexed some of the best minds of the period. In fact, Trembley considered Locke's philosophy so important that late in his life he translated *An Essay concerning Human Understanding* into French. The manuscript of this translation is in the George Trembley archives.

That Locke was a significant element of the Genevan intellectual milieu from which Trembley emerged is demonstrated by Bonnet's reference in his *Mémoires autobiographiques* to a society of friends which he had joined who were reading Locke's *Essay* with "pen in hand." Bonnet reported that their metaphysical discussions nearly put him to sleep. Not only did he deplore the time spent on such abstruse subjects as what constituted the faculties of the soul, but he told his friends that "they would learn more truths during a quarter hour spent at my microscope than in discoursing for months on substances and attributes."[94] As we shall see in the following chapter, the enigmatic characteristics of the polyp soon

---

[93]"*Ame*. Voy ce que pense Mr Locke, sur la Pensée; Il croit que c'est une proprieté que Dieu peut avoir donnée à la Matière. Locke liv. IV chp. 3 & 6. . . . Si Dieu peut donner à la matiere la puissance de penser, ce qu'on apelle l'ame et que l'on conçoit clairement comme indivisible, se pourra pourtant diviser. Car n'y aiant acune partie de matiere qui ne se puisse diviser, je pourrai concevoir que les parties de matiere qui sont en moi, que j'apelle mon ame et qui peuvent penser, peuvent se diviser, je pourrai donc concevoir que mon ame se subdivisera. Cela me paroit un peu absurde." Trembley's Day Book, George Trembley Archives, Toronto, Ontario. Idiosyncratic spelling and punctuation of original text has been retained.

[94]Charles Bonnet, *Mém. aut.*, 61.

drove him to the very metaphysical questions that he had first eschewed. Trembley was never so disparaging of Locke and it is likely that he studied Locke's *Essay* as carefully as he studied natural objects behind the lens of his microscope.

John Yolton's illuminating discussion of the debate over thinking matter forcefully connects Locke's work to British neo-Platonism.[95] Drawing inspiration from Ralph Cudworth's contention that God had added to inert and dead matter certain properties, like sensitivity in some plants and spontaneous motion in the case of animals, Locke had merely extended the notion of providential intervention. God also had the power to add the property of thought to matter.

The debate over the properties of matter was to an extent a reflection of opposition to the Cartesian world machine from which the idea of the animal automaton was deduced. Cudworth and his followers, culminating with Newton, objected to the Cartesian philosophy on religious grounds. The universe was not merely a mechanical orrery set in motion by chance. Forces held the planets as they traced their paths in God's sensorium. No less than the motion of the heavenly bodies, God had brushed living creation with evidence of his providence.

Trembley's appreciation of Locke and his repeated allusions to the polyp's "sensitivity" which he often noted both with respect to its contractility and its response to light must connect him to this tradition of British neo-Platonism which found its way into the natural theology of the late seventeenth century. Whether he was aware of the work of British botanists on plant sensitivity is difficult to ascertain, but it is of interest that not only had they discussed the possibility of "plant animals," but also the idea that "nature makes a scale of creatures." Two of the more well known botanists of this group, Sir Thomas Browne (1605–82) and the microscopist, Henry Power (1623–68) speculated on whether plants might have "a sensation like animals."[96] Browne argued that the idea of plant sensitivity was supported by authorities like

---

[95]See John W. Yolton, *Thinking Matter* (Minneapolis: University of Minnesota Press, 1983).

[96]Henry Power to Reuben Robinson, 2 August 1656, quoted in Charles Webster, "The Recognition of Plant Sensitivity by English Botanists in the Seventeenth Century," *Isi*, 57 (1966): 14.

William Harvey and Pierre Borel. He wrote: "Such a sense may be in plant animals and in the parts of perfect animals even when the head is cutt [sic] off."[97] In addition to his possible awareness of the work in natural history tied to this British neo-Platonic tradition, a more obvious source for Trembley's idea of "plant-animals" and the chain of being is Aristotle, an author whom Trembley read and cited in a variety of contexts.[98]

Whether Leibniz's ideas played a role in the shaping of Trembley's approach to nature is an intriguing question. There is no mention of Leibniz by name, either in Trembley's published work or in his early letters. However, this omission need not indicate that Trembley was completely unaware of Leibniz's ideas. Trembley's study of mathematics, as has already been pointed out, was most certainly strongly influenced by Leibniz's version of the calculus. It seems unlikely that the Genevans, so closely associated with Leibnizians like the Bernoullis, Samuel Koenig and Louis Bourguet, would have been completely insensitive to other aspects of Leibniz's thought, even at a time when his published works were scarce.

The idea of continuity embodied in the chain of being or *scala naturae* was the keystone of Leibnizian metaphysics. Leibniz believed that all of creation, both the organized and the apparently unorganized, was connected by imperceptible degrees, without jumps or lacunae. Animal, vegetable, and mineral realms could not be rigidly separated, as in the Cartesian system. Leibniz had, in fact, predicted that "zoophytes" or "plant animals" would some day be discovered. These plant animals were one of the necessary links connecting the vegetable and animal realms. Leibniz wrote:

Thus there is nothing monstrous in the existence of zoophytes, or plant animals, as Budaeus calls them; on the contrary, it is wholly in keeping with the order of nature that they should exist. And so great is the force of the principle of continuity, to my thinking,

---

[97]Ibid. The quotation is from a letter from Thomas Browne to Henry Power, 8 June 1659, *Works*, 6, 294.

[98]A discussion of Aristotle's system of classification in terms of continuity and his use of sensitivity rather than locomotion as the test for animality can be found in Joseph Schiller, *Physiology and Classification* (Paris: Maloine, 1890), 13. Aristotle is referred to by Trembley, *Mémoires*, 301, though not specifically in connection with the "chain of being" idea.

that not only should I not be surprised to hear that such things had been discovered—creatures which in some of their properties, such as nutrition or reproduction, might pass equally well for animals or plants, and thus overturn the current laws based upon the supposition of a perfect and absolute separation of different orders of coexistent beings which fill the universe;—not only, I say, should I not be surprised to hear that they had been discovered, but, in fact, I am convinced that there must be such creatures, and that natural history will perhaps some day become acquainted with them, when it has further studied that infinity of living things whose small size conceals them from the ordinary observation and which are hidden in the bowels of the earth and the depths of the sea.[99]

Though the letter in which this prediction was expressed was not published until 1753 by Koenig (and was once viewed by scholars as a fake), it does not seem unreasonable to suggest that the Leibnizian characterization of the chain of being was known to Koenig's contemporaries. Koenig, like Cramer, had studied at Basel under Johann I Bernoulli. Nevertheless, the argument for Leibnizian influence on Trembley at this time is tenuous. After Trembley had thoroughly studied the polyp, culminating in the experiment of turning the polyp, he rejected the idea that he had discovered the point of passage between the two realms.

V. THE LEIDEN *MÉMOIRES*

At last in 1744 with the publication of *Mémoires, pour servir à l'histoire d'un genre de polypes d'eau douce à bras en forme de cornes,* Trembley no longer had to balance his intense study of the polyp against his moral responsibility for the instruction of the Bentinck children. His publishers, Jean and Herman Verbeek, had spared no pains to produce a work of the highest quality. Trembley was fortunate in the selection of the Verbeeks because of the extremely liberal conditions that he was able to negotiate.[100] He was allowed as many plates as

---

[99]Translated and quoted by Arthur O. Lovejoy, *The Great Chain of Being,* 145.
[100]Trembley to Bonnet, 24 March 1744, Ms Bonnet 24, Bibliothèque Publique et Universitaire de Genève.

he wished, and they were made by the best engraver; he was able to choose the paper, the type of characters, and the format. The Verbeeks provided him with a stipulated number of copies and four books by other authors.

Only a few months after the publication of the *Mémoires* in Leiden, it was immediately reprinted by the Paris publisher Laurent Durand without Trembley's permission in two small volumes. Some of the magnificent plates of the Leiden edition, on which Trembley and Lyonet had lavished such care, were omitted and it is clear that they were of markedly inferior quality. The pirating of his book did not please Trembley at all because of the liberal conditions and the cordial relations that he had established with the Verbeek brothers. Durand, who also worked for the Imprimerie Royale, presumably proposed to bring out Trembley's book at the same time that Bonnet's appeared, since he had already agreed to publish Bonnet's *Traité d'insectologie*. Thus, Trembley wrote to Bonnet 31 July 1744, "I deeply hope that Durand will not make you wait too long, and that he spares no pains either on the plates or the printing." Trembley reported that Réaumur had written him that the printing of Durand's edition would be even more beautiful than that of the Verbeeks. "That proves," Trembley wrote, "that he can produce beautiful Editions, because mine is beautiful. My publishers have neglected nothing, and that is why it troubles me to see the preparations for pirating their work, because that could do harm to them."[101]

The inferior quality of the woodcuts of the Durand edition in contrast to the copper plate engravings of the Leiden edition was symbolic of the shifting taste of the French reading public after 1745. The observed fact, the intricate details of the life cycles and habits of insects carefully described and illustrated, did not have the same attraction for a new generation of Parisian *philosophes* that it had once had for Réaumur and his followers. In publishing his edition of Trembley's *Mémoires* before Bonnet's *Insectologie* appeared in 1745, Durand (later one of the publishers of the Encyclopédie) may have recognized the new directions that French thought was beginning to take after Trembley's discovery.

By 1744, in addition to Réaumur's announcement of the

[101]Trembley to Bonnet, 31 July 1744, Ms Bonnet 24, Bibliothèque Publique et Universitaire de Genève. (See below, 236.)

discovery in his *Histoire des insectes,* an English amateur, Henry Baker, had also described the polyp in detail in *An Attempt Toward a Natural History of the Polype,* published the previous year. Although Baker gave Trembley credit for the discovery and showed himself a perceptive observer in his own right, that his account was largely based on Trembley's letters to Martin Folkes, caused the president of the Royal Society a certain amount of embarrassment. Trembley's communications to the Royal Society were also first published in the *Philosophical Transactions* in 1743.

Why Trembley did not complete his work more expeditiously raises interesting questions. His uncertainty over the problems of nutrition and reproduction obviously played a role in his lamentable slowness in getting his work to his publisher; but Trembley's extremely careful study of structure, which established the reality of budding in the polyp, and illuminated how the polyp took food, may also have owed a great deal to his concern that his discovery not be misinterpreted. This could have been a far greater motivation than Réaumur's encouragement and example.

Although Trembley, in naming his book *Mémoires,* paid homage to Réaumur's great work on insects, the letters demonstrate that Réaumur generally did not suggest particular experiments to Trembley. Except in the case of eggs, Réaumur's own observations on the polyp did little to influence the direction of Trembley's research. Rather than counseling more extensive study of the polyp, Réaumur urged Trembley to publish. It was Trembley's own extremely cautious attitude toward the problems of the polyp and the care with which he conducted his experiments and which he lavished on his illustrations, that prevented him from being the first into print. If, as Margaret Jacob suggests, Trembley's social milieu at The Hague connected him to a group of Masonic free thinkers who could easily have made the connection between the polyp's regeneration and the radical doctrines of pantheism and materialism,[102] he may have been loath to add fuel to their alchemical fires. Indeed, the 1741 report in the proceedings of the Paris Academy of Sciences might have given

---

[102]Margaret C. Jacob, *The Radical Enlightenment,* 200.

him pause (quoted above p. 7). Its author called regeneration of the polyp more marvelous than mystical palingenesis. Were the symbols of the phoenix rising from its ashes and the rejoining of the serpent simply metaphorical allusions to the hermetic past? Whether Trembley's concern that his discovery of regeneration might be misinterpreted drove him to study his little aquatic being with such fierce discipline and precison cannot be documented. There is nothing in the early letters to Bonnet and Réaumur that illuminates the answers to these questions, except his consistent resistance to the mixing of the observation of nature with the subjective imaginings of the perceiver. If Gabriel Cramer in Geneva had warned against the "fertile imagination" of Descartes, the ideas of Freemasonry were potentially even more dangerous to Christian orthodoxy. Perhaps his most eloquent response to those who might use his work to venture into the realm of materialistic speculation was the final paragraph in his *Mémoires*:

In order to extend our knowledge of natural history, we ought to direct our efforts to the discovery of the greatest possible number of facts. If we knew all the facts contained in nature, we should possess the explanation of them—we should see the Whole that they combine to produce. The more we get to know of them, the more we shall be in a position to examine thoroughly some parts of this Whole. We cannot therefore strive better to explain the facts we know, than by trying to discover new ones. Nature should be explained by nature, not by our own views; for these are too limited to encompass such a great object in its whole extent. The beauty of nature certainly appears more clearly when what we know of it is not mixed with our fancies. So the beauty of nature gives us ideas that are more worthy of the infinite wisdom of its Author, and therefore better suited to shape the mind and heart, than are our fancies. This is what we should hold before us in all our researches.[103]

---

[103]"Nous devons, afin d'étendre nos connoissances sur l'Histoire Naturelle, faire nos efforts, pour découvrir le plus de Faits, qu'il nous sera possible. Si nous connoissions tous les Faits que la Nature renferme, nous en aurions l'explication, nous verrions le Tout qu'ils forment ensemble. Plus nous en connoitrons, plus nous serons en état d'approfondir quelques Parties de ce Tout. Nous ne pouvons donc mieux travailler à expliquer les Faits que nous connoissons, qu'en tâchant d'en découvrir de nouveaux. La Nature doit être expliquée par la Nature, & non par nos propres vûes, qui sont trop bornées pour envisager un si grand Objet dans toute son étendue. La Beauté de la Nature paroit certainement davantage, quand ce que nous en

Clearly, Trembley owed a great intellectual debt to the empirical orientation of the science of 'sGravesande and Boerhaave and he was pleased to have his book published in Leiden. However, if the Netherlands and the Bentinck circle, in particular, were the cradle of a radical materialism that would soon spread to France, this must have strengthened the resolve of this sober Genevan to avoid dangerous speculation. Jacob has suggested that Prosper Marchand (1678–1756), no patient observer of nature's phenomena, but a book dealer, journalist and Freemason, may have acted as "more than a disinterested agent" in arranging for publication of Trembley's book.[104] She has also pointed out that Trembley was acquainted with another Mason, Isaac Sacrelaire. Indirectly he can also be tied to Bernard Picart (1673–1733), one of Europe's master engravers and member of the Masonic order, the Knights of Jubilation. Although Picart had died by the time Trembley was looking for someone to engrave his illustrations, he chose Jacob van der Schley (1715–1779), whom Trembley described as one of Picart's disciples. Nevertheless, it is clear that he based his choice of an engraver on competence, not ideology. Trembley's letters to Bonnet indicate that he was extremely concerned that his illustrations should be of the highest quality and he believed that, though expensive, there were no finer engravers in Europe than in Holland.

Because of the metaphysical questions engendered by the polyp, the discovery was of far greater significance than that of parthenogenesis, which, nevertheless, opened the way for Trembley's dramatic revelations. Bonnet was keenly aware that his painstaking study of his solitary aphid was eclipsed by the polyp's intriguing reproduction. Bonnet immediately recognized the disturbing implications of the discovery for the doctrine of animal soul. The philosophical issues, which he had scorned only a short time before when he read Locke, became the focus of all his subsequent work. In drawing out

connoissons, n'est pas mêlé avec nos Imaginations. Elle nous donne alors des idées plus dignes de la Sagesse infinie de son Auteur, & par cela même plus propre à former l'esprit & le coeur. C'est ce que nous devons nous proposer dans toutes nos Recherches." Trembley, *Mémoires*, 312. Translation by John Baker, *Abraham Trembley*, 240.
[104]Margaret C. Jacob, *The Radical Enlightenment*, 200.

these implications, Bonnet's situation as a student at the Academy in Geneva put him in a position to reflect. His professors were readily accessible and he was at a formative period of his life. A mere twenty years old, though he respected authority, he was not afraid to ask questions. In Holland Trembley had William Bentinck and Lyonet with whom to share his perplexities, and the occasional visits of his friend Allamand. However, contact with the intellectual community at Geneva also appears to have been important to him. Aside from a few letters to Cramer and members of his immediate family, his Genevan connection was maintained through his correspondence with Bonnet. The letters which Trembley and Bonnet exchanged with each other, and which will be discussed in the following chapters, reveal intriguing similarities in their experiments. For this reason, the letters of Bonnet after 1741 become increasingly interesting from the historical point of view, inasmuch as they document the emerging metaphysical crisis of the eighteenth century and Bonnet's attempt to chart a path through those troubled years.

Chapter 5

Bonnet's Response to the Discovery and the Worm With Two Tails

In contrast to their letters to Réaumur, the letters which Bonnet and Trembley exchanged with each other were informal, open and enthusiastic. They are particularly useful to the historian because they were not burdened with the excessive empirical detail that Réaumur expected. The letters between the Genevans were usually short and to the point. They reveal the aspects of the problems which each regarded as important and are useful as a guide through the labyrinth of observations in their published works. The ease with which the two Genevans shared their thoughts shows that each assumed that the other had much of the same background to bring to what was reported by letter.

Trembley's letters to Bonnet reveal that during the course of his studies on the polyp the subject of parthenogenesis continued to hold his attention. The increasingly conclusive evidence that aphids could produce their young "without coupling" reinforced his view that polyps, like aphids, could reproduce without the joining of the two sexes. The discussions of parthenogenesis reveal that even after Bonnet had won the distinction of correspondent of the Paris Academy of Sciences in August of 1740 for his study of aphid reproduction, Trembley was not convinced that all the questions about aphid reproduction had been answered. Parthenogenesis had a liberating effect on Trembley's study of natural history. It made him wary of precipitous conclusions. It was after raising the possibilities that one coupling might serve for several generations, or that aphids might couple in the belly of the mother—the questions that Bonnet claimed led to the loss of his eyesight—that Trembley in his letter of 27 January 1741 gave a cursory description of the perplexing "little aquatic Being" which he had seen for the first time in

137

June 1740. His initial description of the polyp to Bonnet, like the one he gave to Réaumur, revealed his dilemma over whether the little organized body in question should be called animal or plant.

I hardly know if I should call plant or animal the object which occupies me most at present. I have been studying it since the month of June. It has furnished me with quite marked characteristics of both plant and animal. It is a little aquatic Being. Seeing it for the first time, you would exclaim that it is a little plant. But if it is a plant, it is sensitive and ambulant, and if it is an animal, it can grow from cuttings like several plants. I have cut it in three parts. Whatever part each lacked grew back to resemble the Being before it was divided, and each walked and has made up to now all the movements which I have seen the complete animal make.[1]

Bonnet's enthusiastic response is very different from the caution and hesitation that Réaumur had displayed in his early letters to Trembley. It is, indeed, of significance that Bonnet's first reaction, given in his letter of 24 March 1741, was that Trembley had discovered the point that connected the animal and plant realms. Bonnet wrote:

Your little aquatic being is something which is so singular and so surprising that it seems to me that it ought to be regarded as one of the greatest marvels that the Study of Natural History can offer. One can say that you have discovered the point of passage from the Vegetable to the Animal.[2]

Bonnet told Trembley that he was walking in the steps of the "great Swammerdam" and he was delighted that the polyp had escaped Swammerdam's notice, reserving the discovery for a student of Réaumur. Chiding Trembley for the lack of detail in his description, Bonnet revealed that he had shown Trembley's letter to "our professors" (Cramer and Calandrini) and that they were "confounded" by the enigma that Trembley had reported.

But I would have wished that you had spoken a word on the shape of the animal in question, if it is one, its size, color and the place where it is found and other things of this nature which can aid in

[1]Trembley to Bonnet, 27 January 1741, Ms Bonnet 24, Bibliothèque Publique et Universitaire de Genève. (See below, 198.) The full texts of the letters of Trembley to Bonnet and Bonnet's responses discussed here can be found in Appendix A.

[2]Bonnet to Trembley, 24 March 1741, George Trembley Archives, Toronto, Ontario. (Below, 199.)

recognizing it; instead, what you have reported is close to an enigma which I do not know how to decipher. But I easily console myself when I see that the clever men here, even learned men, such as our Professors to whom I have been very pleased to show your letter, are confounded.[3]

Receiving no satisfactory explanation from his professors in Geneva, Bonnet had naturally written to Réaumur for his opinion. "I await with impatience to find out what Mr de Réaumur thinks of your aquatic Being, having asked him to kindly inform me of it.[4] No doubt you should be able to send him a specimen," he wrote. In the same letter Bonnet returned to the question of the reproduction of aphids. He emphatically denied that aphids could couple in the belly of the mother, ending with the caveat: "Let us amass always many facts; then there will be time enough to imagine."[5]

When Trembley responded to Bonnet's letter on 5 May 1741 he informed Bonnet of Réaumur's opinion that the little aquatic being was indeed an animal. "I call it an animal because it is now decided that it is one. It is the opinion of Mr de Réaumur, whom I have been successful in having some reach in a healthy state. He has given them the name of polyp."[6] He told Bonnet of his satisfaction in finding about twelve days previously another species of polyps (hydra vulgaris) that were reddish in color and larger than the first green species. Trembley was pleased because the larger animals would be easier to work with. At this point Trembley had still not fully understood the function of the tentacles, which he described as legs. In addition, the description shows that he had the anterior and posterior ends reversed.

Therefore these are animals, and of a kind which if cut in two or three parts, each part becomes a complete animal. I will explain. Here is the sketch of the polyp.

---

[3]Ibid.
[4]Ibid. (Below, 200.) It is unfortunate that both the letters of 11 March, in which Bonnet presumably asked for Réaumur's opinion of the new discovery, and 19 April 1741 are missing. When Réaumur did respond to Bonnet's letters on 30 April 1741, there is no mention of Trembley's perplexing aquatic being.
[5]Ibid. (Below, 201.)
[6]Trembley to Bonnet, 5 May 1741, Ms Bonnet 24, Bibliothèque Publique et Universitaire de Genève. (Below, 202.)

The threads, a, are its legs. If one cuts the polyps through some portion of the body, b, the part which keeps the threads lives, walks and performs all the functions of the complete polyp, and the legs grow back to the one which doesn't have any at the end of 6, 7, 8, 9, 10 etc., days; and then it performs all the functions of a complete polyp. It is one in every respect. The same thing happened when I cut one in three parts.[7]

In discussing his discovery of the polyp's natural mode of reproduction by budding, he emphasized that no egg was involved. Obviously, he was specifically thinking of Réaumur's suggestion in his letter of 25 March that perhaps Trembley had overlooked a concealed egg.

Another special feature of these animals is their manner of multiplying. The young come from the body of the parent the way the branches come out of a trunk. At first only a little excrescence can be seen which grows each day; then the legs appear and at the end of some time, when the animal is complete, it detaches itself from the body of the mother. I have one from which five are emerging at one time. It is impossible for me to give you a detailed description, but I can clearly prove that in fact the young one comes out of the body of the mother; and that it is not at all like some insects whose eggs or young are attached to the body.[8]

It was apparent that even with a full description of Trembley's little aquatic animal, the problem of the polyp was not resolved for the intellectual community of Geneva. For Bonnet, Gabriel Cramer, whom he regarded as his "oracle," was the logical person to seek enlightenment on the metaphysical quandary that the discovery produced in him. Bonnet's reaction was *not* worry over eggs like Réaumur, however, but consternation over the problem of animal soul. He wrote to Professor Cramer:

---

[7]Ibid. (Below, 203.)
[8]Ibid.

Here is the animal of my cousin which is soundly established! Will we allow it a soul or not? I will dispense with it whenever it pleases you, being the last person in the world in a position to satisfy myself. All that I ardently wish is that my poor insects will not be degraded too much, and I have reason to fear it terribly. I beg you, Sir, strive not to let them become simple machines. I will be inconsolable. Really, I will not observe them with as much pleasure. Good-by then to all industry, all skill, all kinds of intelligence. And if you cannot get them out of this difficulty, who can?[9]

Cramer responded that while Trembley's discovery did seem "to deal a heavy blow to the System of soul in animals," he should not despair.[10] It was too early to draw conclusions.

To Bonnet the discovery seemed a victory for the Cartesian view of the animal as a soulless machine. His letter demonstrates that it caused him considerable anguish. If one could cut the polyp up into twenty or thirty pieces, or even chop it up, and as many new polyps would be regenerated, how could there be one directing soul? Was it divisible? Or were there many souls contained in the body of a single polyp? Why, when the functional unity of the animal machine was destroyed by cutting, did the polyp not die, as might have been expected of other animals? If animals with regenerative properties were granted a soul, why not still lower forms of life? These implications had already been seen by Trembley in a letter he had written to Réaumur three months earlier in March 1741: "Perhaps no time will be lost in stirring up the question whether plants have a soul."[11]

Trembley's discovery of the polyp produced in Bonnet a shift in his own research interests. Though his studies of

---

[9]"Voilà, donc, Monsieur, l'animal de mon Cousin bien constaté! Lui donnerons nous une ame, ne lui en donnerons nous point? J'en passerai partout où il vous plaira, n'etant nullement en état de me satisfaire moi même le moins du monde. Tout ce qu'il y a c'est que je voudrois bien qu'on ne degradat pas trop mes pauvres Insectes; et j'ai terriblement lieu de le craindre. Je vous en suplie, Monsieur, tachés qu'ils ne deviennent pas de simples machines. J'en serois inconsolable. Reellement je ne les observois pas avec autant de plaisir. Adieu alors toute industrie, toute adresse, toute espece d'intelligence. Et si vous ne les tirez d'affaire qui les en tirera?" Maurice Trembley's modernized transcription, 60 (note) in *Corresp. inédite* differs slightly from the original text. Bonnet to Cramer, 29 June 1741, Ms Suppl. 384, Bibliothèque Publique et Universitaire de Genève.

[10]Cramer to Bonnet, June 1741, Ms Bonnet 43, Bibliothèque Publique et Universitaire de Genève. For a transcription of a long passage from this letter, see 156, note 2.

[11]M. Trembley, *Correspondance inédite*, 60.

caterpillars continued to be reported in his letters to Réaumur, Bonnet's new work focused on the phenomenon of regeneration. Unable to find any polyps in the waters around Geneva, in June Bonnet selected a species of aquatic worm, which he suspected shared the same regenerative properties as the polyp. It is unfortunate that several of Bonnet's letters to Réaumur from this period are missing, including the one in which he reported the regeneration of his aquatic worm. Réaumur penned an enthusiastic response to Bonnet's latest revelations 7 August 1741.

I thank you, Sir, for already having verified a prediction which I had made to the Academy, and which could be made without looking like a prophet. Everywhere I have found facts which prove that the Author of nature has multiplied his most singular productions, that He has not limited himself to furnishing us with unique examples of only a few. As soon as it is established that it is very true that a polyp cut in two becomes two polyps, it must be concluded that this strange prerogative has been accorded to other animals, perhaps many others. I suspect that sea nettles which resemble polyps by their horns and by the slowness of their gait also could have it. I recall some observations which appear to prove that starfish do too. Finally, your very curious observations, done with all the intelligence and attention which one could desire, prove incontestably that there is a species of insect of a very different genre from that of polyps which can be multiplied by a means which would be sure to destroy the individuals of other species.[12]

Bonnet's observations, Réaumur reported, had been read in their entirety to the Academy and had been received with great pleasure. Réaumur encouraged Bonnet to extend his

---

[12]"[J]e vous remercie, Monsieur, de ce que vous avez deja verifié une prediction que j'avois faite a l'Academie et qu'on pouvoit lui faire sans se donner pour prophete. partout j'ai trouvé des faits qui prouvent que l'autheur de la nature a multiplié ses productions les plus singulieres, qu'il ne s'est pas borné a nous donner des exemples uniques de quelques unes. des qu'on s'est convaincu qu'il est tres reel, qu'un polype coupe en deux devient deux polipes, on a du conclure que cette etrange prerogative avoit ete accordée a d'autres animaux et peutetre a beaucoup d'autres. je soupconne que ces orties de mer qui resemblent aux polypes par leur cornes et par la lenteur de leur marche, peuvent l'avoir. je me rappelle des observations qui paroissent prouver que des etoiles de mer l'ont aussi. enfin vos observations curieuses, faites avec toute l'intelligence et l'attention qu'on peut desirer prouvent incontestablement qu'il y a une espece d'insectes d'un genre tres different de celui des polipes qui peut etre multipliée par la voye la plus sure pour detruire les individus des autres especes." Réaumur to Bonnet, 7 August 1741, Ms Bonnet 42, Bibliothèque Publique et Universitaire de Genève.

experiments to all "insects of the vermicular form" and suggested that Cramer would be the ideal witness to verify these experiments.

On 4 November Bonnet described his latest "striking" experiments on worms to Réaumur. This time he had divided his worms not only in half, but in three, four, eight, ten and fourteen portions, nearly all of which grew back a head and tail. "So that, Sir, at the hour that I have the honor of writing to you, the majority have made such great progress that there are practically none which cannot be divided easily into ten portions and provide in a few days as many complete animals." Bonnet observed that not all the segments of his worms grew at the same rate, but that it seemed that the segments closest to the original tail grew the most slowly.

Bonnet reflected that plants can multiply by cuttings so perfectly that "the least twig can become a plant." Since God had willed that some insects, his worms for example, have this property as well, his experiments sought to establish whether the Creator had granted this property to worms in the same degree as to plants. Bonnet cut off the heads and the tails of his worms and these parts promptly died. The trunk of the worm continued to live and, in fact, "continued to move as if I had not performed the operation at all; I even saw it bury itself several moments afterwards in the mud, using its anterior end like a Head to blaze a trail there—something which seemed to me extremely remarkable."[13] This experiment, which he repeated several times to make sure that he had in fact not observed incorrectly, led him to the following questions:

Where then does the principle of life reside in such Worms, if after having cut off their heads, they still demonstrate the same movements, I say! Why do they make the same bending motions? But

---

[13]"Parmi les Plantes qui peuvent être multiplées par boutures il y en a en qui cette proprieté semble resider d'une manière plus parfaite, dont le moindre brin peut lui même devenir une Plante. L'Auteur de la Nature qui a voulu que certains Insectes, comme nos Vers, ressemblassent à ces Plantes par cette proprieté, la leur auroit-il accordè au même degré? . . . A égard du Corps il continua de se mouvoir à peu près comme si je n'eu point fait l'operation; je le vis même, ce qui me parut extremement remarquable, s'enfoncer quelques momens après dans la bouë en se servent de son extremité anterieure comme d'une Tête pour s'y frayer un chemin." Bonnet to Réaumur, 4 November 1741, Papers of Réaumur, Bonnet (DB), Archives de l'Académie des Sciences, Paris, 3.

what is this difficulty in comparison with so many others which present themselves to the mind? Are these worms only simple machines: Or are they Composites in which the soul makes their springs move? And if they have in them such a principle, how can this principle find itself in each portion? Will one admit that there are as many souls in these Worms as there are portions of these same Worms which can themselves become complete Worms? Will one believe with Malpighi that these Sorts of Insects are from one end to the other only Heart and Head? All this could be, or at least the impossibility of it is not demonstrated. But in the end do we know more than before? Certainly it is necessary always to return to admire and keep quiet.[14]

Réaumur answered Bonnet's letter immediately. On 30 November 1741, he wrote enthusiastically:

The strangest, Sir, and the most troubling novelty that has ever been offered to those who study Nature is surely the reproduction of animals by cuttings, but as soon as it had been proved that there was a species which could be multiplied by an avenue which is so extraordinary, it must be believed that this species is not the only one to which has been accorded such a surprising property.[15]

Bonnet had been the first to verify experimentally Réaumur's prediction that other examples of regeneration would be discovered. Réaumur's response demonstrated his characteristic approach to natural history. Exceptions to general rules were only appearances. Once a presumed novelty was uncovered,

---

[14]"Ou rèside donc le principe de vie dans de tels Vers. Si après leur avoir coupé la tête ils montrent encore les mêmes mouvemens. que dis-je! les mêmes inclinations? Mais quest-ce que cette difficulté en comparaison de tant d'autres qui s'offrent tout à coup à l'Esprit? Ces Vers ne sont-ils que de simples machines? ou sont-ce des Composés dont une Ame fasse mouvoir les ressorts? Et s'ils ont en eux un tel principe comment ce principe peut-il se trouver ensuite dans chaque portion? Admettra-t-on qu'il y a autant d'Ames dans ces Vers qu'il y a de portions de ces mêmes Vers qui peuvent elles mêmes devenir des Vers complets? Croira-t-on avec Malpighi que ces Sortes d'Insectes ne sont d'un bout à l'autre que Coeur et que Cerveau? Tout cela peut être, ou du moins l'impossibilité n'en est-elle pas demontrée. Mais au fond en est-on plus avancé? Certainement il en faudra toujours revenir à admirer et à se taire." Ibid. The idea that the insect's heart extended from head to tail comes from Malpighi's *Dissertatio epistolica de bombyce* (1669).

[15]"[L]a plus etrange, Monsieur, et la plus embarassante nouveaute qui se soit jamais offerte a ceux qui etudient la nature est assurement la production des animaux par boutures. mais des qu'il a ete prouvé qu'il y en avoit une espece qui pouvoit etre multipliée par une voye si extraordinare, on a du croire que cette espece n'etoit pas la seule a laquelle, une si etonnante proprieté eut ete accordée." Réaumur to Bonnet, 30 November 1741, Ms Bonnet 42, Bibliothèque Publique et Universitaire de Genève.

other examples would be found to demonstrate Nature's underlying regularity. Nevertheless, Réaumur did not answer Bonnet's queries on the question of animal soul in this letter.

Bonnet expanded his observations on worms in a carefully wrought *Mémoire* called *"Sur les insectes qui peuvent être multipliés pour ainsi dire par boutures."* It is dated 7 March 1742 and can be found included among the letters that Bonnet addressed to Réaumur and are preserved in the collection of Réaumur's papers in the Paris Academy of Sciences. This work of twelve pages contains fourteen separate parts in which Bonnet discussed the structure of the worms, their voluntary movements and their responses to various stimuli. These observations were later expanded and published in Bonnet's *Traité d'insectologie.*

Bonnet observed that even after being cut, the halfworm without a head continued to move as though it had one. Moreover, these movements were not like the involuntary movements of the tails of lizards after they had been cut away from the body. The movements of the portions of worms were willed: "therefore, the principle [of life] did not appear to have been destroyed at all."[16] He extended his experiments by cutting his worms into smaller parts, and thereby increasing the number of portions of worms. He was able to divide two worms: one into twenty-four parts, the other into twenty-six parts. Of the worm cut into twenty-four portions, eighteen regenerated their missing parts; while of the twenty-six divisions in the other worm, only nine survived. Comparing his work to the *Vegetable Statics* of Stephen Hales (tr. Buffon), Bonnet attempted to keep track of the rates of growth of his worm segments. With meticulous attention, he observed that the tail sections grew less rapidly. Rates of growth were also affected by seasonal variations of temperature and the presence of food.

---

[16]"[M]ais ce qui me parut bien autrement Remarquable, c'est que l'autre moitié qui n'avoit point de tête se mouvoit presque comme si elle en eut eu une; elle alloit en avant en s'apuyant sur l'extrêmité antérieure de son corps, elle faisoit même Chemin avec assez de vitesse. On voyoit que ce n'étoit point un mouvement sans direction, un mouvement produit par une cause telle que celle qui fait mouvoir la queue d'un lézard après qu'elle a été separée du Tronc; mais un mouvement très volontaire donc le principe ne paroissoit point avoir eté détruit." *Mémoire*, 7 March 1742, Papers of Réaumur, Bonnet (DB), Archives de l'Académie des Sciences, Paris, 4.

In Section VII of his *Mémoire,* Bonnet repeated exactly the
same questions about animal soul which he had already made
in his letter of 4 November 1741. The major differences
between this discussion and the previous one is the inclusion
of the sentence: "Is the marvelous reproduction of all their
parts only a consequence of the laws of motion, or rather
does it depend on a line of germs?"[17]

Bonnet's letter of 5 November 1742 shows that he continued to agonize over the problem. Once more his need to
discuss metaphysical questions was a response to the aston-
ishing results of his continuing experiments on worms. Bonnet had been able to produce experimentally a creature with
no head, only two tails, yet the worm was able to live and to
some extent carry out its life functions, though he recognized
that the creature's life would be limited by its inability to take
food. He was absolutely sure that the new tail, which had
grown in place of the head, was a tail in every respect. He
could clearly observe the anus and he stated that the tail could
perform none of the movements which were characteristic
of a head. It did not contract and stretch itself out, and could
not assist the animal in crawling as the head did. Bonnet
wrote that, in fact, the worm seemed to sense the predicament
that it was in. "It seemed quite abashed *(décontenancé).* However, something which I must not neglect to remark upon,
the Flow of Blood did not seem to have changed direction
at all: it continued to circulate from the posterior end to the
anterior end."[18]

In this letter Bonnet felt justified in going further than
before in his metaphysical queries, since he had more evidence of the serious anomalies which he was discovering with
his experiments on his worms. After repeating his experiments several times, he concluded that his worms with two
tails were not chance monstrosities, since he could create
them experimentally by sectioning them carefully exactly in

---

[17]"[L]a reproduction merveilleuse de toutes leurs parties n'est elle qu'une suite
des lois du mouvement, ou dépend elle plutôt d'une file de germes?" Ibid., 8.

[18]"On sentoit, pour ainsi dire, à son air son embarras. Il sembloit tout deconte-
nancé. Cependant, et c'est ce que je ne dois pas negliger de faire remarquer, le
Cours du Sang ne paroissoit point avoir changé de direction: Il continuoit à circuler
du bout posterieur au bout anterieur." Bonnet to Réaumur, 5 November 1742,
Papers of Réaumur, Bonnet (DB), Archives de l'Académie des Sciences, Paris.

half. Thus, regeneration must depend on a line of germs, a
rational design that demonstrated the creative hand of God.
The crux of the matter rested on the location of the soul.
How could such a "noble principle" be found in a tail? On
what did the selfhood of the insect depend?

What should I think now, Sir, of such an extraordinary fact which
I already seen twice and which I am on the eve of seeing perhaps
once more? Have I surprised Nature making a mistake, so to speak?
Is this one of its monstrous products which sometimes happen
either in the animal or in the vegetable realm? But do such products
show so much regularity? In admitting, as it is rather difficult not
to, that the marvelous reproduction of all the parts of these Insects
is accomplished by a line of Germs placed according to a design,
would chance have decreed that in the worms in question, or more
exactly, in one of the portions of one of these worms—a germ of
tail would have developed rather than a germ of head? Finally,
where does the soul reside, the *self*, in this portion which instead
of a head has regained a tail? It would be very strange that such a
noble principle would be found lodged in a part which is so insignificant. But this is not among the questions to which one can ever
promise to have enlightenment.[19]

If the hapless worm with two tails could still exhibit actions
that were purposeful, was the soul then lodged in the tail as
well as the head? Implicit in the problem of soul raised by
Bonnet was his difficulty with the Cartesian view of matter
as passive, inert, and soulless. If the animal were granted a
non-extended, non-material soul on which the selfhood of
the worm depended, could this "noble principle" be lodged
in the "insignificant" tail? Bonnet may have sensed the direction in which his experiments were taking him, but he did

---

[19]"Que dois-je donc penser maintenant, Monsieur, d'un fait si extraordinaire revû
dèja deux fois et que je suis a la vielle de revoir peut étre encore? Aurois-je surpris
pour ainsi dire la Nature en defaut? Serois-ce ici une de ces productions mon-
strueuses qui s'offrent quelquefois soit dans le Règne animal, soit dans le Vegetal?
Mais de telles productions affectent-elles tant de regularité? En admettant, comme
il est assés difficile de s'en dispenser, que la reproduction merveilleuse de toutes les
parties de ces Insectes se fait par une suite de Germes disposés à dessein, le Hazard
auroit-il voulu que dans les Vers dont il s'agit, ou plus exactement, dans une des
portions d'un de ces vers un germe de queuë se fut developpé plutôt qu'un germe
de tête? Enfin, où reside l'ame, le *Moi,* dans cette portion qui au lieu d'une tête a
repris une derriere: Il seroit bien étrange qu'un principe si noble se trouvat logé
dans une partie qui l'est si peu. Mais une pareille question n'est pas de celles qu'on
peut se promettre de voir jamais éclairer." Ibid.

not allow himself to speculate whether matter might have the power to organize itself without the necessity of a directing soul.

On 21 December 1742 Réaumur responded to Bonnet's letter that contained the anguished questions over animal soul. At the same time he enclosed a copy of Volume Six of the *Histoire des insectes.* This was presumably the first time that Bonnet had seen the introduction in which Réaumur himself tackled the problem of soul. Naturally, by this time Bonnet's observations of the headless worm had carried him well beyond the questions he had originally addressed to Réaumur. Though Réaumur had proposed to publish a ten-volume work on insects, this was the last volume of the series that he completed. It is possible that the problems of a metaphysical nature which the discovery of the polyp raised may have affected the direction of his scientific endeavors which continued for the last fifteen years of his life in new directions related to the problems of generation and digestion in birds. There remain in the Paris Academy of Sciences papers relating to the history of the polyp, which Réaumur may have intended to publish. The content of these notes are tantalizing for any historian determined to clarify Réaumur's philosophical position more fully.[20]

What Réaumur revealed in his discussion in the introduction to Volume Six of the *Histoire des insectes* is his own perplexity and characteristic caution. After discussing the discovery of the polyp by Trembley, he wrote: "These new productions of nature will perhaps serve to give us some information on this mystery of Nature, the generation of animals, which is so hidden and so interesting for us."[21] Regeneration, he affirmed, was a demonstration of the development of germs, the growth of which could be followed from the moment the embryo was visible. But while regeneration could provide new light on embryology, it was more difficult to answer the metaphysical questions which the new discovery engendered. The doctrine of animal soul had become enormously complicated by the new discovery.

---

[20]See "Inventaire de la correspondance et des papiers de Réaumur conservés aux Archives de l'Académie des Sciences de Paris," in *La Vie et l'oeuvre de Réaumur,* 15.
[21]Réaumur, *Histoire des insectes* 6: lxvii.

An interior feeling, and even a sort of spirit of justice causes the majority of men not to know how to bring themselves to refuse a soul to animals: few Philosophers believe it justified to treat them as pure machines; but are there souls which are divisible? What sorts of souls would those be which like the body would let themselves be cut into pieces and would reproduce themselves again? If the soul in animals has an assumed place in which it keeps itself in the manner of souls, and if this place is in the head, will we imagine that each piece of the body is not only provided at its anterior end with a germ of the head, but that in addition this germ of head contains within it one of the soul; that is to say, to the germ appropriate to become a head, is attached a soul which will not be in a state to exercise its functions until the germ of the head has developed, has acquired the power to carry on the functions of the head, and has become that of an animal?[22]

By couching his discussion in the form of questions, Réaumur avoided an issue that he rightfully regarded as metaphysical and ultimately not answerable on the level of natural science. However, it is worthy of note that his queries imply that he believed that the head must be the seat of the soul, a Cartesian presupposition which he was loath to give up.

In the letter that accompanied the gift of Volume Six, Réaumur did Bonnet the honor of responding to the specific issue of a worm with two tails, though his answer again demonstrated his skepticism of these strange new facts so characteristic of him. However, his statement revealed his belief that there were, throughout the bodies of animals that regenerate, germs of the anterior and posterior parts which must be touching. The germs do not develop until conditions are favorable for their growth. Réaumur, ever cautious, suggested that it remained to be discovered what these conditions

---

[22]"Un sentiment intérieur, & même une espece d'esprit de justice, font que le commun des hommes ne scauroit se résoudre à refuser une ame aux animaux: peu de Philosophes se croyent fondés à les traiter de pures machines; mais y a-t-il des ames secables? Quelles sortes d'ames seroient-ce que celles qui, comme les corps, se laisseroient couper par morceaux & se reproduiroient de même? Si l'ame dans les bêtes a un lieu affecté, où elle se tient à la manière des ames, si ce lieu est dans la tête, imaginerons-nous que chaque tronçon du corps est non seulement pourvû à son bout antérieur d'un germe de tête, mais que de plus ce germe de tête en contient un d'ame; c'est à dire qu'au germe propre à devenir une tête, est attachée une ame qui ne sera en état d'exercer ses fonctions, que quand le germe de tête se sera développé, qu'il aura acquis la puissance de faire les fonctions de tête, & qu'il sera devenu celle d'animal?" Ibid.

were which could stimulate the development of a tail instead
of a head.

These observations of tails which are born where heads ought to
be born are extremely unusual, and I do not despair that you will
succeed in repeating them more than once. When the fact is well
established, the difficulty will not be to find the germ in the posterior part which has been produced, for it is necessary that there
are germs of the anterior and posterior parts throughout these
animals which are touching, and some are caused to develop in
preference to others only when the end where it is located is the
most favorable to their development. It remains to be discovered
what can in some circumstances facilitate the development of a
posterior part on an anterior end, that is to say the one closest to
the head.[23]

Réaumur's thinking on this issue is not entirely clear. He
appears to have thought that by design God had spread
throughout the body of the animal germs of the posterior
and anterior parts which were contiguous. Réaumur asked
Bonnet to determine the external or environmental conditions that might stimulate a tail to grow instead of a head.
He avoided the question of what is the unifying principle of
the organism, the self *(le moi)*, as Bonnet called it. This question later became one of the central themes of Bonnet's philosophy.[24] The problem of the unifying principle revealed
the weakness of the position of those who accepted the Cartesian distinction between extended matter and non-extended
soul, yet were determined to allow soul in animals. If animals
were granted a soul, this soul must provide the unity on which
selfhood depended. However, to grant soul to specific parts

---

[23]"Ces observations de queues qui sont nées ou des tetes devoient naitre sont
extremement singulieres, et je ne desespere pas qu'il ne vous arrive de les refaire
plus d'une fois. le fait etant bien constaté l'embarras ne sera pas de trouver les
germes de la partie posterieure qui a ete produite, car il faut qu'il y ait partout dans
ces animaux des germes de partie anterieure et de partie posterieure qui les touch-
ent, et les unes ne sont determinées a les developper preferablement aux autres
que lorsque le bout ou elles se trouvent est le plus favorable a leur developpement
restera a scavoir ce qui peut en quelques circonstances faciliter le developpement
d'une partie posterieure sur un bout anterieur, j'appelle ainsi, le plus proche de la
teste." Réaumur to Bonnet, 21 December 1742, Ms Bonnet 42, Bibliothèque Publique et Universitaire de Genève.

[24]This question is broached on a philosophical level for the first time by Bonnet
in *L'essai de psychologie* (1755). See also Lorin Anderson, *Charles Bonnet and the Order
of the Known*, 105 ff. See Jean Starobinski, "L'essai de psychologie de Charles Bonnet:
Une version corrigée inédite." *Gesnerus* 32 (1975):1–15.

of the animal led to logical difficulties. On what then did the functional unity of the organism depend?

Bonnet's extremely long letter of 7 January 1743 began with an enthusiastic acknowledgment of the receipt of Volume Six of Réaumur's *Histoire des insectes*. However, Bonnet did not mention the issue of animal soul in this letter. It appears that the problem was again raised in either his letter of 30 August or 18 September (both of which are unfortunately missing) because Réaumur returned to a discussion of germs in his letter of 10 November. In his discussion of Bonnet's observations on worms with two tails, Réaumur hinted that Bonnet had not taken pains enough to observe with care. Perhaps, the observations themselves, which led to such surprising conclusions, were flawed. Had a new tail grown in place of the head, or had Bonnet mistaken a tail for a head?

Is it certain that the newly reproduced tail is a tail like the old one? could it be suspected that it is only a head which is more tapered than ordinary? for the structure of the head is not always perceptible enough to be distinctly seen in those which are very slender.[25]

Sometimes when cuttings were made the parts grew so slowly that it was difficult to perceive whether they were heads or tails. However, if indeed, there was no mistake, then Réaumur would admit that the multiple germ theory was the most plausible explanation. "These tails which grow back in place of the head seem to me to be very favorable to the proposed opinion that the new and strange reproductions of cut-up insects are due to germs whose development is occasioned by the cutting."[26] Evidently, Bonnet had also included new observations on worms that produced two heads and no tail in his missing letters. These, Réaumur cautioned, needed to be repeated, but seemed to indicate that the old worm was in the process of giving birth to its offspring when it was sectioned artificially by Bonnet. The "tubercules" or buds emerging from the body of the mother would have eventually separated from the parent if Bonnet's cutting of the worm had not interfered with it. This might explain why Bonnet

[25]Réaumur to Bonnet, 10 November 1743, Ms Bonnet 42, Bibliothèque Publique et Universitaire de Genève.
[26]Ibid.

had observed that each headed section appeared to behave with an independent will.

Because of the loss of Bonnet's subsequent letters to Réaumur (with the exception of two letters of minor importance written in 1746), it is not possible to continue to explore Bonnet's views on the problem of animal soul in the Bonnet-Réaumur correspondence. However, it is clear that the issue was of paramount importance to Bonnet and that it did not cease to provide grist for his developing philosophical interests. This fact can be clearly seen in the letters which he exchanged with his mentor Gabriel Cramer in Geneva.

From Réaumur's responses to Bonnet's queries it appears that Réaumur believed that germs held within the bodies of animals were passive until external conditions triggered their development. Germs were conceived by Réaumur to be kinds of eggs or embryos that contained preformed within them those parts capable of regeneration. Réaumur does not elucidate, but it appears that he thought that in addition to the soul-carrying germs encased within the ovary, there must be additional specialized germs which stimulated the growth of a new part when the original had been severed from the body of the animal. This explanation was consistent with the one which he had proposed in 1712 for the regeneration of the claws of crayfish in his article in the *Histoire et mémoires de l'Académie Royale des Sciences*, "Sur les diverses reproductions qui se font dans les écrevisses, les omars, les crabes, &c., & entr'autres sur celles de leurs jambes & de leurs écailles."

All that we might propose as being the most convenient and most reasonable would be to suppose that these little legs which we see born were each enclosed in little eggs, and having cut a part of the leg, the same juices which serve to nourish and to cause to grow this part are employed in causing the development and birth of the kind of little germ of leg enclosed in this egg.[27]

---

[27]"Tout ce que nous pourrions avancer & de plus commode, & peut-être de plus raisonnable, ce seroit de supposer que ces petites jambes que nous voyons naître, étoient chacune renfermées dans de petits oeufs, & qu'ayant coupé une partie de la jambe, les mêmes sucs qui serroient à nourrir & faire croître cette partie, sont employés à faire développer & naître l'espèce de petit germe de jambe renfermé dans cet oeuf." *Histoire de l'Académie Royale des Sciences*, 1712 (Paris: Pankoucke, 1777), 307–308. See also Jean Rostand, "Réaumur embryologiste et généticien" in *La Vie et l'oeuvre de Réaumur* (Paris: Centre International de Synthèse, Presses Universitaires de France, 1962), 99–115; Jacques Roger, *Les Sciences de la vie dans la pensée française du XVIIIᵉ siècle* (Paris: Armand Colin, 1963), 392–94.

Réaumur proposed this explanation only as a "supposition," and he had been careful to review the problems which such an explanation might encounter. However, while the explanation had an admittedly *ad hoc* quality, Réaumur implied that he saw no better explanation. His explanation of the regeneration of the legs of a crayfish was an extension of the preexistence argument. He applied the same principles to the reproduction of parts of living bodies that he believed operated in the generation of complete organic machines.

Even at the time of Réaumur's publication of his study of the regeneration of the legs of the crayfish, his contemporaries were skeptical. As Fontenelle pointed out in response to Réaumur's report in the History of the Academy for the year 1712, regeneration of parts of individuals "is a second marvel of a different nature" from that of whole organisms, which is not explained by the theory of eggs.[28] According to ovist theory, the egg was the vehicle for the preformed miniature. The soul, contained within the germ, provided the unity of the animal on which its selfhood depended. But if there were specialized germs to facilitate the regrowth of lost appendages, were the germs for the regeneration of the legs of crustaceans soulless?

While the explanation of the regeneration of crustacean appendages put a strain on the preexistence explanation, Bonnet's experiments on worms introduced even greater perplexities that could not be dismissed. On the preexistence model, the design of the insect was unalterably built in. The fixity of the species and the mechanical nature of the animal body demanded that the patterns on which it was predicated could not vary. Bonnet's experiments demonstrated that design could be affected experimentally, yet the behavior of the animal continued as though the structure could be altered and its functional unity interfered with, without destroying the animal's capacity for life.

Bonnet's question where the noble principle, the life principle, was located was in fact a challenge to the basic Cartesian idea of the animal machine. Bonnet's questions on the problem of soul could only be resolved by adopting a new concept

---

[28]Alain F. Corcos, "Fontenelle and the Problem of Generation," *Journal of the History of Biology* 4 (1971):370.

of matter—no longer passive and inert in the Cartesian view, but dynamic, intrinsically possessing the capacity for life.

Though the debate over animal soul cannot be traced further in the Bonnet-Réaumur correspondence because of the loss of Bonnet's later letters, it is possible to continue to probe Bonnet's thoughts in the period prior to 1749 through his letters to his compatriots Trembley and Cramer. The letters illuminate the points on which Bonnet and Trembley seemed in agreement and how each contributed to the formulation of new questions of generation and animal soul.

Chapter 6

Metaphysical Debate in Geneva

Though Bonnet raised the issue of animal soul in his letters to Réaumur at least three times, Réaumur chose not to respond directly by letter to his questions. This was not the case with Cramer, who carefully answered Bonnet point by point. Though Cramer and Bonnet were not in daily personal contact at the Academy, on the days when Bonnet did not have regular classes with Cramer, he was free to correspond with him. He appears to have taken full advantage of this opportunity and on the days that he remained at the family's country house at Thonnex, a few miles outside Geneva, he shared his thoughts with his professor by letter. Often Bonnet directed Cramer to have his courier leave his response with Bonnet's grandmother Lullin who lived in the city. Bonnet also shared the letters he received from Réaumur with his professor.

The Bonnet-Réaumur exchange showed that despite Réaumur's lack of response, by 1742 Bonnet had arrived at a positive statement about animal soul, which he suggested depended on a line of germs placed by God's design in the body of the insect. Bonnet arrived at this position not through Réaumur's influence, but through his role in the metaphysical debate in Geneva over the implications of the discovery of the polyp. In this debate Bonnet occupied the center stage and took on the role of transmitter of the Genevan interpretation to Trembley.

When Bonnet wrote his anguished letter to Cramer on 29 June 1741[1] expressing his fear that the metaphysical consequences of Trembley's discovery would be the abandonment of the idea of animal soul, Cramer answered immediately. Although Cramer could not reassure Bonnet that the final resolution of the problem would not be a victory for the

---

[1] Quoted above, 141.

Cartesian animal automaton, he attempted to assist Bonnet
by referring him to further reading:

In reading what you said, I remembered several articles of Father
Pardies on the understanding of animals. Look at the place which
I marked. As he cites St. Augustine, he should be consulted. Is it
for me to enlighten you concerning something about the nature
of Insects which you know so perfectly! It is true that your Observations are a severe blow to the System of soul of animals. But
nevertheless I will not say that all is to be despaired of yet. A learned
ignorance could well be the limit and the conclusion of our rea-
sonings. In the last resort if they are machines, they are all the
more admirable. For you only accord to them a soul because with
it you [can] more easily explain their actions which appear free and
purposive. If it were well proven that they have no soul at all, it
would be necessary to renounce this explanation which is so simple
and so natural. Things would not happen with as much ease as we
believe. They would be only more admirable. But say good-by to
their industry, their skill. No, it is only necessary to ascend higher,
and this industry, this skill will be that of the Author of these
Machines. But I see clearly that you are attached to insects on
account of the ability to feel that you believe you see in them, and
that your ethical sense [*le moral*] makes more touching a pleasure
which seems completely physical. Well! Let us hope that it is not
impossible for them to keep the ability to feel.[2]

[2]The text of Cramer's letter to Bonnet (with the exception of the closing sentence)
is as follows: "Je ne sais, mon cher Monsieur, comment vous y prenez; mais il me
semble que vos Observations deviennent de jour en jour plus interessantes. J'ai lu
votre lettre avec toute l'attention qu'elle mérite & j'en ai d'abord été bien rencom-
pensé par le plaisir que cette lecture m'a donné, plaisir qui a duré jusqu'au para-
graphe où je vois que vous avés été malade. Au nom de Dieu, mon cher Monsieur,
conservés-vous, Menagés une santé devenue précieuse au genre humain. La perte
de votre vers, m'auroit chagrinée, si je [ne la] regardois pas comme un coup de la
providence qui veut fixer un peu votre attention sur votre santé, pour vous mettre
en Etat de pousser dans la suite l'histoire naturelle au dela des bornes où nous la
voions réduite. Je me suis rapellé, en vous lisant, quelques articles du P. Pardies de
la connoissance des bètes. Voiés l'endroit où j'ai mis une Marque. Comme il cite Sᵗ
Augustin, il faudra le consulter. Est ce à moi à vous éclairer en quelque chose sur
la Nature des Insectes que vous connoissés parfaitement! Il est vrai que vos observations portent un rude coup au Systeme de l'ame des bètes. Mais je ne dirai pourtant
pas encore que tout soit désespéré. Une docte ignorance pourroit bien ètre le terme
& la conclusion de nos raisonemens. Au pis aller, si ce sont des Machines, elles n'en
sont que plus admirables. Car vous ne leur donnés une ame, que parce que vous
expliqués plus aisement par là leurs actions, qui paroissent libres & tendre à certaines
fins. S'il croit bien prouvé qu'elles n'ont point d'ame, il faudroit renoncer à cette
explication si simple et si naturelle. [L]es choses ne se passeroient plus avec autant
de facilité que nous croions. Elles n'en seroient donc que plus admirables. Mais,
dites-vous, adieu leur industrie, leur adresse. Non, il faut seulement remonter plus
haut & cette industrie, cette adresse sera celle de l'Autheur de ces Machines. Mais

No doubt Bonnet did consult the passage in Pardies's *Discours de la connoissance des bestes* which Cramer had marked for his study. Pardies, as already pointed out, accepted the idea of animal soul. Though a Cartesian in other respects, he took the neo-scholastic or peripatetic position that the soul of animals is composed of a third substance, between brute matter and spirit. This sensitive soul, while material, endowed the animal with sense perception and the ability to distinguish between simple objects.[3] It accounted for its spontaneous behavior, which derived from the animal's ability to express emotions. Though deprived of reason, their possession of *"connoissances sensibles"* made animals closer to man than the Cartesian position allowed.

Though a chain of being is not actually stated by Pardies, it is implied by his stress on the basic similarity between animal emotions and those of humans. Pardies denied that there was a specific location of animal soul; it could not be located in either the head (Descartes) or in the heart (Aristotle), but must be "spread throughout the body, & is not indivisible and unique."[4] It is in his discussion of the need to conceive of the soul as divisible that Pardies refers to St. Augustine:

But I ask you, let us examine a little how this can be understood & let us consider one of these little animals with several legs [centipede] similar to the one about which St. Augustine spoke in the book on the Quantity of Soul. The saintly doctor recounted that one of his friends took one of these animals which he put on a table, & that he cut it in two; & that at the same time these two parts which had been cut began to walk, & to flee very quickly, one to one side, and the other to the other. This was not an irregular movement; they walked with the same exactness demonstrated by the entire animal: When something was placed in their path, or they were hit on one side, they ably turned aside, & fled toward the other side. Once more each of the parts was cut, & there ap-

---

je vois bien, que vous vous affectionés aux insectes par les sentimens que vous croiez apercevoir en eux, & que le moral rend plus piquant un plaisir qui semble tout physique. Eh, bien! esperons qu'il n'est pas impossible que le sentiment ne leur reste." Cramer to Bonnet, June 1741, Ms Bonnet 43, Bibliothèque Publique et Universitaire de Genève. Since this letter appears to be a response to Bonnet's letter of 29 June, the date may not be correct.

[3]Gaston Ignace Pardies, *Discours de la connoissance des bestes 1672* (New York: Johnson Repr. Corp., 1972), 147ff.

[4]Ibid., 77. Actually, Descartes located the soul in the pituitary gland of the brain.

peared then four pieces which walked, as if there had been four different animals; & although they were divided still again, each little piece still lived. . . .

St. Augustine said that this experiment enraptured him, & that he remained for some time without knowing what to think about the nature of the soul. And in fact, if we suppose that the soul of these Animals has the faculty of feeling and of perceiving, as we feel and perceive; certainly what is shown by this experiment, will be not only admirable, but incomprehensible.[5]

Pardies, following St. Augustine, then asked what became of the sense of self *(le moi)* that has the ability to feel pain and notice objects, once the insect is divided? Can the self be in two places? Shouldn't it be considered indivisible?[6]

The behavior which Bonnet observed in his worms was similar to that of the sections of St. Augustine's centipede. The movements of the two segments of a worm cut in half also seemed voluntary, and they could respond to various stimuli. He observed that each half of a worm cut in two could turn aside at the encounter of an obstacle in its path, stop, then resume crawling. When exposed to strong sunlight, they crawled faster, and appeared startled when touched lightly with a stick.[7] Though these observations comprise Bonnet's letter to Réaumur of 4 November 1741, it is obvious from his allusions to these experiments in the shorter letters to Cramer and Trembley, that he was working on them

---

[5]"Mais je vous prie examinons un peu comment cela se peut entendre, & considerons un de ces petits Animaux à plusieurs pieds, semblable à celui dont parle Saint Augustin au livre de la Quantité de l'ame. Ce saint Docteur raconte qu'un de ses amis prit un de ces Animaux, qu'il le mit sur une table, & qu'il le coupa en deux; & qu'en même temps ces deux parties ainsi coupées se mirentà marcher, & à fuir fort vîte, l'une d'une côté, & l'autre de l'autre. Ce n'étoit pas un mouvement irrégulier; elles marchoient avec la même justesse qu'auroit fait l'animal entier: Lors qu'on leur opposoit quelque chose, ou qu'on les frapoit d'une côté, elles se détournoient fort bien, & s'enfuïoient vers un autre endroit. On coupa derechef chacune de ces parties, & il parut pour lors quatre piéces qui marchoient, comme si c'eût été quatre animaux différens; & quoi qu'on les partageât encore davantage, chaque petit morceau vivoit encore. . . .

S. Augustin dit, que cette experience le ravit en admiration, & qu'il demeura quelque temps sans sçavoir que penser de la nature de l'ame. Et en effet, si nous supposons que l'ame de ces Animaux ait la faculté de sentir, & d'appercevoir, comme nous sentons, & comme nous appercevons; certainement ce qui se voit dans cette experience, sera non seulement admirable, mais incomprehensible." Ibid., 78–79; 80–81.

[6]Ibid., 82–84.

[7]Bonnet to Réaumur, 4 November 1741, Papers of Réaumur, Bonnet (DB), Archives de l'Académie des Sciences, Paris.

during the summer of 1741. It was typical of Bonnet to wait
to write to Réaumur until he had thoroughly examined a
particular research problem.

The question of animal soul continued to plague Bonnet.
A letter to Cramer revealed that he had consulted his other
professor, Calandrini, to assist him with an explanation which
would account for these signs of soul in the behavior of his
mutilated worm segments. Bonnet proposed Calandrini's explanation to resolve the metaphysical crisis in a letter to Cramer, 1 August 1741:

You are aware, Sir, of the little System which Professor Calandrini
has imagined to conserve a soul in each portion of our worms. My
first experiment was not very favorable to it, as you can remember;
I have done others which have proved to me that immediately after
the operation each portion has movements and inclinations nearly
as before. I have seen some bury themselves in the silt.

But to limit myself to the reproduction of the head or the tail,
it seems to me that it is hardly less difficult to explain. MR Bourguet
with whom I have had the honor of dining at the home of the
Reverend Derochemont, imagined for this as far as we could understand it, germs of heads and of tails arranged in a line one after
the other, the whole length of the body of the insect.[8]

But, Bonnet implied that until he had his professor's opinion
on the matter, it was just a hypothesis.

Bonnet's intriguing reference to Calandrini's system cannot be explored without further research into unpublished
sources, for Calandrini published little. Calandrini's germs,
debated over dinner with the venerable Leibnizian from Neuchâtel, Louis Bourguet, is significant, for it establishes the
connection between the Genevans and the philosophy of
Leibniz. Bourguet, known as the "Pliny of Neuchâtel," was a

---

[8]"Vous savés, Monsieur, le petit Systeme qu'a imaginé M^r le Prof^r Calandrini pour
conserver une ame à chaque portion de nos Vers. Ma 1ere experience ne lui étoit
déjà pas trop favorable, comme vous pouvés vous ressouvenir; j'en ai fait d'autres
qui m'ont prouvé qu'immediatement après l'operation chaque portion a des mouvemens et des inclinations presque comme auparavant. J'en ai vû s'enfoncer dans le
limon.

Mais à se borner à la reproduction de la tête et de la queuë il me semble qu'elle
n'est gueres moins difficile á expliquer. Mr Bourguet avec qui j'ai eu l'honneur de
manger chéz le Pasteur Derochemont imaginoit pour cela, autant nous le pûmes
comprendre, des germes de têtes et de queuës disposés a la file les uns des autres
tout du long du corps de l'Insecte." Bonnet to Cramer, 1 August 1741, Ms Suppl.
384, Bibliothèque Publique et Universitaire de Genève.

close friend of Cramer and Calandrini. Their joint venture
the *Bibliothèque Italique,* published in Geneva between 1728
and 1734 stressed to its French-speaking readers the importance of the experimental work of Malpighi and Vallisnieri.
Since Bonnet reported in his *Mémoires autobiographiques* that
he and Calandrini learned to do dissections on insects using
Malpighi as a text, it would seem that Bourguet was successful
in his mission to have the Italian work recognized. Bourguet
had not only corresponded with Leibniz, but also had popularized his ideas in his *Lettres philosophiques,* published in
1729 in Geneva and reviewed in the *Bibliothèque Italique.*[9]
Though Bonnet refers here to Bourguet's ideas in a slightly
disparaging tone, Bourguet's ideas were important in the
formulation of Bonnet's later metaphysical ideas, particularly
those found in the *Considérations sur les corps organisés.*[10] This
early personal contact with Bourguet the year before Bourguet's death seems to have been extremely important in Bonnet's intellectual development.

From the little that Bonnet tells Cramer in this letter, it
appears that Calandrini's system may have provided Bonnet
with an alternative to the soulless automaton of Descartes
and yet avoided the passive *emboîtement* idea of Malebranche.
Germs spread throughout the body of the polyp, each of
which contained soul-like properties, seems to have offered
an alternative to the view of eggs encased one within the
other as the passive vehicle for "life" or soul. It avoided the
logical difficulties that resulted from the need to envision the
possibility of infinitely small germs and contradicted the
Cartesian idea that an animal had merely a functional unity
that was living only so long as the parts worked together
according to mechanical laws.

Whether Calandrini's germs were similar to Leibnizian
monads cannot be determined in any absolute sense without
more knowledge of Calandrini's thought. However, it seems
plausible to suggest that at this time, prior to his reading of
the *Theodicy,* through the influence of Calandrini and Bour-

---

[9]For an excellent article on Bourguet by François Ellenberger, see *Dictionary of
Scientific Biography,* ed. Charles Gillispie (New York: Charles Scribners, 1978), 15,
Suppl. I: 52–59.

[10]Joseph Schiller, "La Notion d'organization dans l'oeuvre de Louis Bourguet
(1678–1742)," *Gesnerus* 32 (1975): 94.

guet, the importance of the Leibnizian ideas of matter and animal soul were beginning to take shape in Bonnet's mind. Leibniz had believed that matter was dynamic and was organized at the Creation. However, to impose unity on this mass of living matter in order to create an organized body, a soul was necessary. Leibniz declared:

In the beginning when I had liberated myself from the yoke of Aristotle, I was drawn to the void and Atoms, for this is what satisfies the imagination best. But having come back to them after a great deal of meditation, I perceived that it is impossible to find the principles of a true Unity in matter alone or in what is only passive, since all is only a collection or mass of parts to infinity."[11]

True unity in the organized body could only be guaranteed by the soul.

Calandrini's system continued to intrigue Bonnet, for he returned to it in a letter to Trembley in September 1741. It had been a busy summer, full of debate among the Genevans over the implications of Trembley's discovery. Bonnet had many experiments on his worms behind him, and his view of how to explain the new phenomenon was begining to take shape. He wrote that he had found Trembley's experiment of cutting the polyp in sections lengthwise among the hardest to explain because such sectioning should have destroyed the "Animal Oeconomy."[12] Bonnet pointed out that while the behavior of the polyp suggested a soul, its manner of reproduction by sectioning led to problems that needed solution on a philosophical level. Calandrini's system was suggested as a way out, though Bonnet's allusion to the torment that Trembley's organized body had caused the Genevan intellectual community shows that the Genevans may not have been entirely satisfied with Calandrini's germ theory as a solution to the enigma which the polyp presented.

The industry which you have noticed in your Polyp will no doubt not give a great deal of pleasure to the Metaphysicans: if on one side it seems to prove that it has a soul, on the other side its ex-

---

[11]Quoted from a letter to Jaquelot, 22 March 1703 by Jacques Roger, "Leibniz et les sciences de la vie," *Akten des internationalen Leibniz Kongresses* (Wiesbaden, 1969) 2: 211.

[12]Bonnet to Trembley, 1 September 1741, George Trembley Archives, Toronto, Ontario. For text of letter, see Appendix A, 206.

traordinary reproduction gives birth to terrible difficulties. Will
there be in this Insect as in those which resemble it in its manner
of reproduction, as many souls as there are portions of these same
Insects which can themselves become perfect Insects? Do these souls
in this state still only have a simple idea of their existence; can they
only perform their functions when the operation has given stimulus
to the germs in which they are enclosed to develop themselves?
This is in large [part] the system of Mr. Calandrini. I do not know
how it will seem to you, but I doubt that several of our gentlemen
will agree with my observations. Immediately after the operation,
I saw one of the portions of one of my worms, the one which had
only kept its tail, act like the one which had kept its head. I have
performed other observations whose success has been the same or
nearly.

Your organic body torments us; we would be very grateful to
you if you would be willing to speak about it at a little greater
length. Will there be other animals to whom Nature has given the
ability to multiply itself in the manner of certain plants?[13]

It is not clear to whom Bonnet is referring in his remark
about the possibility that the discovery will not give a great
deal of pleasure to Geneva's metaphysicians. It may be a
reference to the strict Cartesians, who denied any animal soul
at all. The behavior of both polyps and worms immediately
after sectioning indicated a measure of voluntary activity or
soul. The Cartesians would have expected a sectioned animal
to die. Professor Calandrini's germs offered an explanation
for why an animal deprived of its functional unity still retained its life principle. Bonnet's abbreviated discussion of
the theory suggests that Trembley was already familiar with
Calandrini's system.

However, since this reference to the metaphysicians is obscure, another possible explanation should be offered. Bonnet may be referring to the supporters of neo-Platonism, who
believed in a system of controlling intelligences. While the
polyp's ability to regenerate might seem to support their
interpretation of animating forces spread through the matter
that composed the body of the polyp, the "terrible difficulties," in this case, would be the need to account for the providential action of God in the creation of these beings. The

---

[13]Ibid.

fact that these perplexing implications were clearly perceived by Bonnet can be seen in his letter to Professor Cramer, written several months later on 17 August 1742:

I would greatly wish to know, but I fear, Sir, that I am abusing your kindness, if the experiment which I am doing of cutting from the same worm the head and the tail as fast as they grow back, seems to you as curious as it seems to me. It could at least furnish some enlightenment on several interesting questions to which I have referred in my letter to MR de Réaumur of last June 23. Today I am making the 7th cut. The Partisans of the system of *Intelligences rectrices,* that of Mr. Hartsoeker, would they not believe themselves well supported by a similar experiment?[14]

Nicholas Hartsoeker's (1656–1725) system of controlling intelligences, could provide an explanation of animal soul derived from the plastic forces of Neo-platonism, but such a view of matter might also open the door to the proponents of epigenesis. Bonnet did not actually make this connection here, but it was certainly one of the problems that the Genevans must have anticipated. Several years later, when Bonnet began a correspondence with Caspar Cuentz (or Kunz), known as the metaphysician of St. Gall, he was well prepared. Cuentz was the author of a four-volume work, *Essai d'un sisteme nouveau concernant la nature des êtres spirituels, fondé en partie sur les principes du célèbre M. Locke, philosophe anglois, dont l'auteur fait l'apologie* (Neuchâtel, 1742). Their letters discuss Cuentz's objections to the idea of multiple souls in the bodies of polyps which Cuentz had argued were lazy and perished with the death of the animal. Cuentz also proposed an epigenetic theory of generation, his system of *"homoeomères,"* which Cramer vigorously opposed. He wrote in 1745:

---

[14]"Je souhaiterois bien de scavoir, mais je crains, Monsieur, d'abuser de vôtre complaisance, si l'expérience que je fais de couper à un même par la tête et la queuë à mesure qu'elles lui reviennent vous paroît aussi curieuse qu'elle me le semble. Elle peut fournir au moins des eclaircissemens à plusieurs questions interessantes dont j'ai cotté quelques unes dans ma Lettre à M. de R. de 23 Juin dernier. Je fais aujourdhuy la 7e coup. Les partisans du systeme de *Intelligences rectrices,* du M. Hartsoêker, ne seroient-ils pas cru[?] bien appuyés par une semblable experience?" Bonnet to Cramer, 17 August 1742, Ms Suppl. 384, Bibliothèque Publique et Universitaire de Genève. Though I have the letter of June 23 to Réaumur, one page is missing and I was not able to find the passage to which Bonnet refers here.

The head of the most vile insect is a machine which is more com-plicated *(composée)*, more perfect, more beautiful than our most admirable clocks. When I have seen a clock formed by pure mech-anism, with *homoeomères*, I could begin to believe that a head could be formed in the same manner.[15]

However, in 1741 these issues concerning animal soul were just beginning to take shape. Trembley and Bonnet both turned to their demanding experimental programs deter-mined not to speculate on the metaphysical problems that their discoveries engendered. Trembley's letter of 22 Decem-ber 1741 shows that he continued the line of reasoning sug-gested by Bonnet, namely, that making cuts in the polyp destroyed its functional unity but not its principle of life. Trembley then described his famous experiment in which he created a "hydra" by sectioning the polyp lengthwise, part way down the body of the polyp, while leaving the partially sectioned pieces attached. Then he cut off the seven "heads" which he had created, repeating on a small scale the "val-orous" exploit of Hercules.

I have some which were born the month of June, and which have not ceased to multiply from a few days after their birth up to the present without coupling. It seemed to me that the manner in which Polyps multiply is extremely analogous to that by which plants multiply by shoots. These singular animals are certainly even more admirable than you can imagine. I made hydras with seven and eight heads, by cutting them lengthwise part way down the body beginning at the head. Then I did the exploit of Hercules. I val-orously cut off the seven heads from one hydra. Seven heads came back, and what is more, each of the seven heads which were cut off will soon be in a state to become a Hydra. For this I am careful to nourish them well. I have a polyp which was already cut in 36 parts and most of these 36 multiply and all would multiply if I wished it.[16]

---

[15]"La tête du plus vil insecte est une machine plus composée, plus parfaite, plus belle que nos horloges les plus admirables. Quand j'aurai vu une pendule formée par le pur méchanisme, avec des homoeomères, je pourrai commencer à croire qu'on peut former de la même manière une tête." Letter of 20 July 1745, quoted in *Mem. Aut.*, 68. Bonnet shared with Cramer his letters from Cuentz and his responses. This exchange is summarized by Jacques Marx in *Charles Bonnet contre le lumières*, 199–205.

[16]Trembley to Bonnet, 22 December 1741. (See below, 207.) Trembley's experi-ments with grafting are discussed by Howard M. Lenhoff and Sylvia G., Lenhoff, "Tissue Grafting in Animals; Its Discovery in 1742 by Abraham Trembley as he Experimented with Hydra," *Biological Bulletin* 166(1984): 1–10.

Bonnet's "abashed" two-tailed worm was an experimental creation similar to Trembley's hydra. He was particularly enthusiastic about this discovery, though the problems that it created on a metaphysical level were particularly tortuous.

You are tempted to suspect that here is some illusion,—some rather imperfect resemblance, such as is sometimes seen in several products either in the animal kingdom or in the vegetable. Not at all. This tail which grew in place of the head is absolutely similar in all parts to the true tail of these Worms. I have assured myself of this by more than one observation. It is, I assure you, something amusing to see the trouble which it causes these poor worms to whom this misfortune has happened. . . . Here is certainly something on which to make conjectures.[17]

Bonnet's remarks in his letters to Trembley reveal that he regarded their experiments as complementing each other. Though each was using a different species of regenerative "insect," their results were strikingly similar. Bonnet took satisfaction in the fact that they did not share the complete details of their experiments. That allowed them to keep their minds more open as they discovered more surprising facts about their ability to create fascinating variations on nature's patterns.

There is no harm done in our not communicating our observations in more detail, [for] the truth will shine to greater advantage. . . . In a word, there is always a gain in questioning nature in several ways; its responses furnish more light.[18]

The possibility of playing with Nature's designs, so well demonstrated by these experiments in grafting would have been impossible on the old Cartesian model in which life in an organized body depends on the exquisite arrangement of parts. Rather, they seem to presuppose a view of matter as dynamic and organized. Both Trembley and Bonnet had been able to demonstrate that infinite variations could be

---

[17]Bonnet to Trembley, 21 November 1742, George Trembley Archives, Toronto, Ontario. (Below, 216.) Bonnet's tissue-grafting experiments are discussed by Gerhard Rudolph, "Les débuts de la transplantation expérimentale—Considérations de Charles Bonnet (1720–1793) sur la 'greffe animale'" in *Gesnerus* 34 (1977): 50–68.

[18]Bonnet to Trembley, 28 June 1742, George Trembley Archives, Toronto, Ontario. (Below, 213.)

produced in animals which possessed the ability to regenerate.

On 11 December 1742 Trembley wrote Bonnet from The Hague, complimenting him on his work on aquatic worms. "Your worm with two tails is admirable, but it does not surprise me, because nothing surprises me."[19] Then he briefly described the most shocking experiment of all—that of turning the polyp inside out.

I have again made several experiments on the Polyps, the details of which would take too much time and space. Their body is a sack which could be compared to a hose-pipe. I have turned it inside out the way one turns a stocking or a glove. The interior became the exterior, and the exterior, the interior, and the Polyps on which I performed this experiment lived, ate, and multiplied.[20]

For Bonnet this experiment was the most important that Trembley had described, and he immediately wrote to Cramer to share with him, in as much detail as he could, all the particulars. In addition, he discussed with Cramer some of the many problems concerning the structure of the polyp which the experiment of inversion posed. He attempted to relate the ability of polyps to turn themselves to other known examples of inversion in nature, but it was difficult to establish a true precedent for the phenomenon. While it was conceivable that parts of animals might turn inside out, he could find nothing analogous to the complete inversion of an animal. He wrote to Cramer in the conversational style which was so uncharacteristic of his letters to Réaumur:

Ah! Please, Sir, what is this? An animal which can be multiplied by cutting, an animal whose young come out of the body as a branch comes from a trunk, finally an animal which is turned inside out like a *stocking* or *glove* and which keeps on *living, eating* and *multiplying.* Here are the marvels, the wonders which we owe to Mr Trembley. Oh, how will we ever sufficiently admire the surprising works of the Creator? What a study is that of Natural History! In truth I have seen nothing like it. But how was Mr T. led to [the idea of] about turning Polyps inside out? How has he managed it? For myself I have difficulty believing that he has inverted them

---

[19]Trembley to Bonnet, 11 December 1742, Ms Bonnet 24, Bibliothèque Publique et Universitaire de Genève. (Below, 217.)

[20]Ibid.

whole. I would believe it more willingly if it were only on portions that he operated. These portions, being nearly everywhere of an equal diameter, sorts of hollow cylinders, the thing appears to me thus more capable of being done. Whatever is the case, our great observer has done this experiment. What can now be said? Is the body of these insects formed of two skins in whose lining or *duplicature* is found all the parts necessary for life? We have examples of analogous inversion. Swammerdam admired what happened to the parts which concern generation in male bees and Mᴿ Réaumur judged it in fact very admirable. Swammerdam also caused us to admire what happens to the heads of snails, to their horns, etc. Finally, sea nettles open up and close; they are sorts of Mouths which invert themselves as you will see, I think, in the Mem. de l'Acad. for 1710 or 11.[21]

In Bonnet's mind, inversion was a bombshell that immediately led him to the resolution of his metaphysical quandary, for it solidified the thought, playing about in his head ever since he had written to Trembley, that the polyp was the "point of passage" linking the animal and vegetable realms.

All this convinces me more and more of the thought which I have that the Polyps, my aquatic worms and all the animals which come from cuttings, are the animals whose structure is the simplest and which are closest to that of Plants. No doubt all is arranged by degrees, by rungs. First the Minerals, 2nd the Vegetables, 3rd the Insects which can be multiplied by cuttings, 4th those which do not

---

[21]"Ah je vous en prie, Monsieur, qu'est-ce que ceci? Un animal qui peut être multipliés de *boûture,* un animal dont les petits sortent du Corps comme une branche sort d'un Tronc, enfin un Animal qu'on retourne ni plus ni moins qu'un *bas* ou qu'un Gant & qui ne laisse pas de *vivre, de manger et de multiplier.* Voilà les merveilles, les prodiges que nous devons à M. Trembley. Eh, comment suffisons nous jamais a admirer les Ouvrages surprenans du Createur! Qu'elle Etude que celle de l'Histoire Naturelle! En verité je n'ai rien vû. Mais comment M. T. a-t-il été conduit à retourner les Polypes? Comment s'y est-il pris? Pour moi j'ai peine à croire qu'il les ait retournés entiers. Je penserois plus volontiers que ce ne sont que des Portions sur lesquels il a operé. Ces portions étant à peu près partout d'un égal diametre, des especes de Cylindres creux, la chose me paroît ainsi plus faisable. Quoiqu'il en soit, nôtre grand Observateur a fait cette Experience. Que dire presentment? Le Corps de ces Insectes seroit-il formé de deux peaux dans la doublure ou duplicature desquelles se trouveroient logées toutes les parties necessaires à la Vie? Nous avons des Exemples d'un retournement analogue. Swammerdam a admiré celui qui arrive aux parties de la generation des males abeilles, et M. Reaumur la jugé en effet tres admirable. Swammerdam nous fait encore admirer celui qui arrive au Cerveau des Limaçons, a leurs Cornes, &c. Enfin les Orties de Mer s'epanouissent, se referment; elles sont des especes de Bouches [?] qui se retournent, comme vous le verrés je crois, dans les Mem. de l'acad. pour 1710 ou 11." Bonnet to Cramer, 21 December 1742, Ms Suppl. 384, Bibliothèque Publique et Universitaire de Genève.

undergo metamorphoses, 5th those which metamorphose, 6th the Reptiles, 7th the Fish, 8th the different species of large animals, terrestrial and winged, finally, Man; here is in large how I would compose my ladder. But what a lot of intermediate degrees! I return to the polyps whose insides have been put on the outside. I do not willingly pardon MR T. for not having told me more about it.

Bonnet ended his letter with a variation of the formal closing which demonstrates his devotion to his teacher: "If men could be thus turned, what would not be seen in the interior of the majority! What would certainly be seen inside me would be my respectful feelings with which I will always be, Sir, your very humble and very obedient servant."[22]

In his response, which lacks a date but which must have been written almost immediately, Cramer supported Bonnet's interpretation of the chain of being, only asking him for further "intermediate steps" between the mineral and vegetable realms. In his discussion of turning the polyp inside out Cramer revealed that he was also familiar with the idea that animals should have interior roots or lacteal vessels. Boerhaave's definition of animals as inside-out plants, which had played such an important role in the discovery of the polyp, had not been mentioned in Trembley's letter to Bonnet describing inversion. Therefore, it appears that Boerhaave's definition was known and accepted by the Genevans.

Let me breathe a little. You are overwhelming us with marvels. Don't you know that in the end your wonders will cease to seem so, if you increase them so greatly. It is annoying, as you say, that MR Trembley is so laconic. Do battle with him and make many reproaches on my part. We don't ask for infinite details, but a little more clarity since what he proposes is an Enigma. Like you, I believe that he has turned inside out some portion of these Worms. For if he had turned them inside out completely, their arms would have

----

[22]"Tout cela me confirme de plus en plus dans la pensée [ou je tend?] que les Polypes, mes Vers aquatiques & tous les Animaux reviennent de bouture, tout [sic, for tous] les animaux dont la structure est la plus simple et qui se raproche le plus de celle des Plantes. Tout va sans doute ici par degrés, par Echellons. D'abord les Mineraux, 2e les Vegetaux, 3e les Insectes qui peuvent [?] être multipliés de bouture, 4e ceux qui n'ont point de metamorphoses a subir, 5e ceux qui se metamorphosent, 6e les Reptiles, 7e les Poissons, 8e les diverses Espèces de grands Animaux terrestres et volatils, Enfin l'Homme; voilà en gros comment je composerois mon Echelle. Mais que de degrés intermediaires! Je reviens aux Polypes dont le dedans a été mis en dehors. Je ne pardonne pas volontiers a M.T. de ne m'en avoir pas dit davantage." Ibid.

been hidden in the interior, unless they were implanted exactly on the edge of the sack. It is necessary for the animal to have lacteal veins (or the equivalent), as much outside as within, since the outside must become the inside. Your examples of analogies are very good, but they are still a long way from the worms of MR Trembley. How is it that you have not been able to find any? You would have derived a great deal from it, and you would have found yet another admirable feature. Your ladder of Beings pleases me a great deal. But I would put some gradations in the Minerals and in the Vegetables, the way you do in the animals. Some minerals are completely crude, such as the *molasses*[?], etc. Others have something which approaches more of an organization, and others draw near them afterwards like the Crystals, the Asbestos, and other fibrous rocks, etc. Among the Vegetables there are some which hardly merit the name. Mushrooms, moss, lichen; others are more perfect, and those which are sensitive, etc. approach the Animal.[23]

After receiving this letter from Cramer, Bonnet wrote to Trembley, specifically asking him if the polyp did not have lacteal veins. His letter, written on the 27th December, shows that he lost no time in exploring the issue of turning the polyp. It was probably with reluctance that he even took time to observe the Christmas festivities.

In truth, Sir and dear Cousin, you push our admiration to the limit. What! you have new Wonders! . . . Here are some Animals which not only can be multiplied by cuttings and whose young emerge

[23]"Laissés moi un peu respirer. Vous nous accablés de prodiges. Savés vous qu'à la fin vos merveilles ne le seront plus, si vous les multipliez si fort. Il est facheux, comme vous dites, que Mr Trembley soit si laconique. Faites lui en la guerre, & bien des reproches de ma part. On ne lui demande pas des details infinis, mais un peu plus de clarté que s'il se proposoit un Enigme. Je crois comme vous qu'il a retourné quelque portion de ces Vers. Car s'il les avoit retourné tout entiers, les bras se seroient cachez dans l'interieur, à moins qu'ils ne soient implantés precisement au bord du sac. Il faut que cet animal ait des Veines laitées (ou l'équivalent) tant au dehors qu'au dedans, puisque le dehors peut devenir le dedans. Vos Exemples analogues sont trés bien, mais il y a encore du chemin de là aux vers de Mr Trembley. Que n'en avés vous pû trouver? vous en auriés tiré grand parti, & vous y trouveriés encore quelque singularité admirable. Votre Echelle des Etres me plait beaucoup. Mais je ferois des degrés dans les Mineraux & les Vegetaux, comme vous en faites dans les animaux. Il est des Mineraux, qui n'ont rien que de fort brute; telles sont les Molasses, &c. D'autres ont quelque chose qui aproche plus d'une organisation, & D'autres en aprochent assés, comme les Crystallisations, les Amianthes & autres Pierres fibreuses &c. Parmi les Vegetaux, il en est qui ne meritent presque pas ce nom. Champignons, Mousse, lichen, D'autres sont plus parfaits, et les Sensitives, &c aprochent de l'Animal." Cramer to Bonnet, no date but almost certainly written between 21 December and 27 December 1742, Ms Bonnet 43, Bibliothèque Publique et Universitaire de Genève.

from the Body as a branch of a tree comes out of a trunk, but
which can even be turned inside out, neither more nor less like a
stocking and thus turned do not stop living, nourishing themselves
and multiplying. You are very malicious in not telling us more and
never speaking to us except in Enigmas. MR Cramer, and all here
who are scholars and honest gentlemen who admire you make the
the same complaint of you. You do not have any conscience at all
to put our minds in torture this way. On my own account, I would
very much like to know how you have been led to perform such
an experiment and how you know how to go about it. I do not
understand how you could try it with success on entire polyps, and
I imagine it better using portions: in truth the structure of these
insects is barely known to me. I would desire also for you to apprise
me whether this marvellous turning inside out happens naturally,
as I am very inclined to believe. . . . Is the Body of these Insects
formed of a double membrane in whose lining Nature has placed
the veins and all the parts which are necessary for them to have
lacteal veins or the equivalent as much outside as within? Finally
do Polyps turned inside out, grow equally fast and multiply like
those which have not been turned? Can they also be multiplied by
cutting, and in all respects?[24]

Trembley answered Bonnet's questions about inversion in
his letter of 8 February 1743. Though he did not allude to
Boerhaave's definition, it is clearly his idea that animals were
reversed plants that led him to the experiment. His greater
detail in the letters to Réaumur has shown Boerhaave's role
clearly, as has been demonstrated in Chapter Four. Trembley
wrote to Bonnet carefully answering each question:

It is the idea, although very vague, which I have of the structure
of the polyps, which made the idea of turning them come to me.
I had already tried it in 1741, but without success. I returned to it
last summer, and after many troublesome attempts, I found a very
short and easy method to use. Polyps never turn themselves inside
out by themselves. It is on a complete Polyp that I do the experiment. It appears that the state in which the polyp finds itself after
having been turned does not please it at all. Nearly always it tries
to turn itself back, and often it succeeds. Ordinarily, I have recourse
to an expedient to prevent the Polyps from turning themselves
back. I threaded a reversed polyp with a pig's bristle, I passed it

---

[24]Bonnet to Trembley, 27 December 1742, George Trembley Archives, Toronto,
Ontario. (See below, 218.)

near its anterior end, near its lips: when they wanted to turn themselves back, they were stopped by the pig bristle. It is nothing for
a polyp to be threaded in this manner. It eats and multiplies as if
it were nothing to have a pig's bristle passing through its body.[25]

There is a break in the correspondence between February
and April because Bonnet was busy studying for his law ex-
ams. However, as soon as he was successful in passing them,
he wrote again to Trembley. He admitted that it was lucky
that he had not found any polyps in the waters around Geneva, for they might have been too distracting. In his letter
of 26 April 1743, inversion was again on his mind and he
shared with Trembley his belief in a "ladder of organized
beings" in which the polyp figured as the animal "simplest
and closest to that of plants."

Well, you are doing the surprising operation of turning entire
polyps inside out, and to keep them from turning back, you pass
through them, near their lips, a pig bristle. You add that it is
nothing for a polyp being thus threaded to eat and multiply as if
nothing had been done. Here is what convinces me more and more
in the feeling of mine that of all the Animals the polyps are among
those whose structure is simplest and closest to that of plants. It
seems to me that if a ladder of organized beings were set up, they
must be placed immediately above the vegetables. But how in this
simplicity these insects are surprising, and how worthy of being
admired is the wisdom of the one who has arrived at unveiling to
us such marvels.[26]

For Bonnet the experiment of turning the polyp inside out
was the confirmation of the chain of being. It provided the
empirical proof of what had previously only been a philosophical notion. Without the grounding in observed fact, it
is doubtful that Bonnet at this point in his career would have
found the idea at all appealing.

Was Bonnet familiar with Leibniz's prediction of the discovery of plant-animals? There is no indication in any of his
writings prior to 1747 that he had any direct contact with
Leibnizian works. Since he later became an enthusiastic sup-

---

[25]Trembley to Bonnet, 8 February 1743, Ms Bonnet 24, Bibliothèque Publique
et Universitaire de Genève. (See below, 219–20.)

[26]Bonnet to Trembley, 26 April 1743, George Trembley Archives, Toronto, Ontario. (See below, 221–22.)

porter of Leibniz, it is likely that he would have mentioned
Leibnizian works in the *Mémoires autobiographiques* if any single work had been particularly influential in his early school-
ing.

Lorin Anderson has written in *Charles Bonnet and the Order
of the Known* that the idea of a chain of being was possibly
"so common in the eighteenth century that any attempt to
trace its source in Bonnet's thought would be impossible."[27]
However, he asks why, if the idea was so common, did Bonnet
take such pride in it? And why did contemporaries attempt
to figure out who had inspired it?[28] Anderson has suggested
that Bonnet may have been led to the idea after reading
Pope's *Essay on Man,* and he points out that in Bonnet's "Preface" to the *Insectologie* there are several references to the
Essay on Man, though there is no specific connection made
by Bonnet between Pope and the idea of the chain. "At the
very least, they show that Bonnet had read the poem and
still had it very much in mind as he composed it."[29] From the
evidence of the letters, it would seem that Trembley's demonstration of turning the polyp inside out confirmed Bonnet's
idea that the polyp was an intermediate form. Pope is never
mentioned.

What is strikingly clear is that the Genevans were excited
over the implications of the polyp for the concept of chain
of being, whereas Réaumur took little interest in these speculations. While Bonnet and Cramer both viewed the polyp
as an intermediate form connecting the two realms, Réaumur
never once in his published work or in his correspondence,
referred to the polyp as an "intermediate being," a "plant-
animal," or a "point of passage."[30] Through the absolute separation of the animal and vegetable realms, Réaumur was
able to preserve the basic Cartesian dualism. It should be
noted, however, that Trembley, though he considered the
idea that the polyp might be a "plant-animal" or zoophyte in
his early letters to Réaumur, when he came to present his

---

[27]Lorin Anderson, *Charles Bonnet and the Order of the Known* (Boston: Reidel, 1982),
6.

[28]Ibid., 7.

[29]Ibid.

[30]See Joseph Schiller, *Physiology and Classification: Historical Relations* (Paris: Maloine, 1980), 125.

conclusions in his *Mémoires,* did not claim that the polyp was the point of passage linking the two realms. Though he and Bonnet shared similar backgrounds and education, it was perhaps his absence from Geneva during these years of crisis, and also his possible awareness of the fairly obvious materialistic implications of his work that kept him from plunging into the metaphysical debate.

Bonnet did not rely on Réaumur's advice when he was contemplating his first published discussion of the chain of being in his introduction to the *Insectologie.* In fact, Réaumur had counseled Bonnet to provide neither notice nor preface.[31] The letters between Bonnet and Cramer show that Bonnet disregarded Réaumur's advice and went ahead with the idea of preparing a general introduction. Bonnet met with Cramer at his house and together they discussed the "difficulties" that Cramer assisted Bonnet in clarifying. There can be no doubt that Cramer had considerable influence over the ideas it finally contained. Bonnet wrote after one of these sessions:

I came away from your home yesterday, quite convinced that it seemed appropriate for me to work on this General Introduction. But as soon as I returned here and I started to think it over, the difficulties immediately reappeared and got the upper hand in the struggle when I was alone. I was vanquished by them, overwhelmed, and I could not recover no matter what effort I made. In this state, Sir, I have the honor of writing to you to ask you to lend me a strong hand. Come to my assistance and all these difficulties which surround me will disappear.[32]

Since the most important idea of Bonnet's introduction is his "Ladder of Natural Beings," it is certain that Cramer played a large role in its conception. When Bonnet shared

---

[31]"Ne mettez à la tête de votre ouvrage ni avertissement, ni préface!" Réaumur to Bonnet, 13 July 1744, Ms Bonnet 26, Bibliothèque Publique et Universitaire de Genève.

[32]"Je sortis hier de chés vous assés persuadé, à ce qu'il me sembloit qu'il convenoit de travailler à cette Introduction Generale. Mais dès que je fus de retour ici, et que je me fus mis à reflechir là dessus, les Difficultés reparurent aussitôt, et combattirent avec d'autant plus davantage que j'etois seul. J'en fus vaîncu, terrassé, et je ne puis me relever quelqu'effort que je fasse. Dans cet état, Monsieur, j'ai l'honneur de vous écrire pour vous prier de me prêter main forte. Venès à mon secours, et toutes difficultés qui m'environnent s'évanouiront." Bonnet to Cramer, 4 December 1743, Ms Suppl. 384, Bibliothèque Publique et Universitaire de Genève.

MAN
Orangoutang
Ape
QUADRUPEDS
Flying squirrel
Bat
Ostrich
BIRDS
Aquatic birds
Amphibious birds
Flying Fish
FISH
Crawling fish
Eels
Water snakes
Slugs
Snails
SHELLFISH
Tubular worms
Moths [or Tinea?]
INSECTS
Gallflies
Tapeworms
Polyps
Sea anemones
Sensitive [plants]
PLANTS
Lichens
Moulds
Mushrooms, agarics
Truffles
Corals, Coraloids
Lithophytes
Asbestos
Talc, gypsum, selenites
Slates
STONES
Figured Stones
Crystallizations
SALTS
Vitriols
METALS
SEMI-METALS
SULPHURS
Bitumens
EARTHS
Pure Earth
WATER
AIR
FIRE
More subtile substances

Tom. I. Pag. Ire.

IDÉE D'UNE ÉCHELLE
DES ETRES NATURELS.

L'HOMME.
Orang-Outang.
Singe.
QUADRUPEDES.
Ecureuil volant.
Chauve-souris.
Autruche.
OISEAUX.
Oiseaux aquatiques.
Oiseaux amphibies.
Poissons volans.
POISSONS.
Poissons rampans.
Anguiles.
Serpens d'eau.
SERPENS.
Limaces.
Limaçons.
COQUILLAGES.
Vers à tuyau.
Teignes.
INSECTES.
Galinsectes.
Tænia, ou Solitaire.
Polypes.
Orties de Mer.
Sensitive.
PLANTES.
Lychens.
Moisissures.
Champignons, Agarics.
Truffes.
Coraux & Coralloides.
Lithophytes.
Asbeste.
Talc, Gyps, Sélénites.
Ardoises.
PIERRES.
Pierres figurées.
Cryftallisations.
SELS.
Vitriols.
METAUX.
DEMI-METAUX.
SOUFRES.
Bitumes.
TERRES.
Terre pure.
EAU.
AIR.
FEU.
Matieres plus fubtiles.

Bonnet's "Idea of a Ladder of Natural Beings" facing page 1 of the *Traité d'Insectologie*.
Permission to reproduce courtesy of Department of Special Collections, Case Western Reserve University Libraries.

with Réaumur the draft of his introduction several months later, Réaumur did not oppose Bonnet's concept of a "ladder of beings," but he criticized it for lacking "the passage of non-organized beings to organized beings, the rocks and the minerals to plants." Moreover, he recommended to Bonnet that he not have it engraved, because that would make what was after all only a hypothesis look too final. "You will not be as obliged to provide an exact enumeration. You will furnish a sketch which you can finish best afterwards."[33] Again, Bonnet chose not to follow Réaumur's advice and had his Ladder engraved on a separate sheet, facing page one. The published result of Bonnet's conversations with Cramer revealed none of his earlier hesitation and ambivalence.

In fact, if we examine the principal productions of Nature we will easily notice that between those of different classes and even those of different genera, there are those which seem to hold a middle place, and form thus as many points of passage or links. This is what can be seen above all in the polyps. The admirable properties which they have in common with Plants, that is to say the multiplication by *budding* and by *cuttings* indicates sufficiently that they are the tie which unites the vegetable Kingdom to the animal. This reflection gave birth to the perhaps foolhardy thought to draw up a Ladder of Natural Beings, which one finds at the end of this Preface. I do not produce it except as an essay, but appropriate to make us conceive of the most lofty ideas of the system of the World and of the Infinite Wisdom which has formed and combined the different pieces. Let us render ourselves attentive to this beautiful spectacle. Let us see the innumerable multitude of organized and non-organized bodies placed one above the other according to the degree of perfection or of excellence which is in each.[34]

---

[33]Réaumur to Bonnet, 30 August 1744, Ms Bonnet 26, Bibliothèque Publique et Universitaire de Genève.

[34]"En effet, si nous parcourons les principales productions de la Nature, nous croirons aisément remarquer qu'entre celles de differentes classes, & même entre celles de différens genres, il en est qui semblent tenir le milieu, & former ainsi comme autant de points de passage ou de liaisons. C'est ce qui se voit sur-tout dans les Polypes. Les admirables propriétés qui leur sont communes avec les Plantes, je veux dire, la multiplication *de Bouture* & celle *par rejettons*, indiquent suffisamment qu'ils sont le lien qui unit le regne végétal à l'animal. Cette réflexion m'a fait naître la pensée, peut-être téméraire, de dresser une Echelle des Etres naturels, qu'on trouve à la fin de cette Préface. Je ne la produis que comme un essai, mais propre à nous faire concevoir les plus grandes idées du systême du Monde & de la Sagesse Infinie qui en a formé & combine les différentes pièces. Rendons nous attentifs à ce beau spectacle. Voyons cette multitude innombrable de corps organisés, & non organisés, se placer les uns au-dessus des autres, suivant le degré de perfection ou d'excellence qui est en chacun." Charles Bonnet, *Oeuvres d'histoire naturelle et de philosphie* (Neuchâtel: Fauche, 1779–83) I:xxx.

Through the underlinings Bonnet emphasized the polyp's unique modes of reproduction by buddings and cuttings. He did not mention the possibility of eggs, though he must have been well aware of Réaumur's discussion of eggs in the introduction to Volume Six of the *Histoire des insectes*, especially as Réaumur had sent him the volume as a gift and referred in this introduction to the animal-soul question, taking some of the very words that Bonnet had used in framing his questions.

For Bonnet the idea of a ladder of creation was a way to resolve the problem of animal soul. There were degrees of soul, just as there were degrees of organization. The view of nature as a continuum meant that animal, vegetable, and mineral could not be rigorously separated since each being shared characteristics in common with those above and below it. The soul was necessary to provide the unity that distinguished one body from another. This unity was not imposed merely by an arrangement of parts but was intrinsic to the matter that composed the body. It seems that, even in this early period, in Bonnet's mind the three concepts of continuity, animal soul, and dynamic matter were related, just as they were related in the Leibnizian system. However, it seems unlikely that Bonnet was consciously using Leibnizian concepts at this stage of his development. Prior to the discovery of the polyp, Bonnet had not been interested in philosophical abstractions, but took delight in his observations. Observations, however, are not carried out in an ideological vacuum. Bonnet had not, as yet, formulated a consistent philosophy of science, and he was influenced by ideas from a variety of writers. However, even before he encountered Leibnizian philosophy directly, the problems with which he was grappling, particularly those on grafting, pointed toward a solution in terms of a Leibnizian theory of matter. The results of these experiments challenged the Cartesian analogy between the clock and the living being. When the wheels and gears of a clock were rearranged, it stopped working, but the worm with two tails, though its days were numbered, continued to live. Trembley's success in producing "hydras" also supported this dynamic theory of matter. Calandrini's "system" seemed to explain the ability of germs to assume the functions of soul after sectioning. Whether Calandrini's germs shared characteristics with Leibnizian monads remains a tantalizing question which cannot be answered before a more thorough investigation of Calandrini has been undertaken.

The role of Louis Bourguet must also be carefully considered before any final answers on the transmission of Leibnizian ideas to the Genevans can be provided. However, the fact that he was invited to Geneva at the crucial point in the metaphysical debate over the implications of the discovery of the polyp does not seem to be coincidental. Bonnet considered this dinner significant enough to report it to Cramer by letter; and the idea of germs of heads and tails of his worms arranged in a line, as the letters to Réaumur over the problem of animal soul demonstrate, continued to hold his attention.

Bonnet made a great deal of the evidence of animal soul shown in the purposive behavior of his insects. It is certain that he was familiar with Pardies and Régnault, both of whom refused to give up the idea of animal soul, though in other respects they accepted the Cartesian philosophy. Pardies had specifically asked the question, attributed to St. Augustine, on what did the selfhood or *le moi* of the insect depend? What was the principle of unity, if after cutting, an insect continued to show signs of purposive activity? Leibniz had also asked this very question, arguing that an animal soul must provide the unifying principle.

Bonnet's interest in the metaphysical issues which the polyp raised, combined with the failure of his eyesight in 1744, caused in part by the counting of successive generations of aphids, ultimately drove him away from the observational natural history of the early correspondence with Réaumur and Trembley. In 1748 he had his first known direct contact with Leibniz through the reading of the *Théodicée*, an experience which he viewed as a decisive influence on his intellectual development.[35] However, Bonnet had been well prepared for his encounter with Leibniz by his involvement with three prominent Leibnizian ideas: the chain of being, animal soul, and dynamic matter. His sensitivity to these ideas was fostered by his careful education under Cramer and Calandrini.

---

[35]Charles Bonnet, *Mém. aut.*, 100.

Chapter 7

Conclusion

The implications of the discovery of the polyp for the philosophy of living beings gradually unfolded in the letters exchanged among Bonnet, Trembley and Réaumur between 1740 and 1745. These letters are a precious record of scientific and human interaction among three very different individuals. Through them the historian is given access to questions and perplexities seldom included in the reports of careful observations and experiments published in their books. These letters reveal the intimacies, the records of conversations over dinner and the web of human connections which stretches beyond the published record. By retaining a freshness and candor, but more important, a trail of ideas and problems often excised from published works, they add some flesh to the dry skeleton of the history of science, so often written as a tiresome record of ever greater discovery.

Admittedly, there are drawbacks to using letters as sources. These include unedited repetitions, occasional missing or torn pages, and notes dashed off without regard for spelling, punctuation or handwriting. However, in this case, it has usually been the very letters which were the most frustrating in a technical sense which have yielded the greatest intellectual treasure. The loss of Bonnet's letters to Réaumur, possibly removed from the archives of the Paris Academy after Réaumur's death by the vindictiveness of his rival, Buffon, remains a serious problem, which makes reconstruction of this exchange more difficult.[1] It is hoped that some of these letters will turn up in private collections. Despite these limitations, through this study of the Bonnet-Réaumur-Trembley correspondence, Réaumur's role as an eighteenth-century patriarch of science is better understood.

Réaumur, a Parisian of established scientific reputation,

---

[1] I have explained the possible reasons for the disappearance of these Bonnet letters in Appendix C, 247.

wealth and important Continental influence, deserves the credit for inspiring Bonnet's discovery of parthenogenesis through suggestions which he made in his widely read *Mémoires pour servir à l'histoire des insectes*. This work was of unquestionable influence in shaping the interests in natural history of an entire generation of investigators, among whom Bonnet and Trembley were the finest. In encouraging the research of Trembley and Bonnet and communicating their discoveries to the Academy of Sciences, Réaumur was fulfilling not only a professional obligation which he owed to the Academy according to its regulations of 1699 concerning correspondents, but also what he regarded as his mission to raise the quality of natural history observations through the rules which he set forth in his introduction to Volume Two.

It is ironic that Bonnet and Trembley, and Trembley's unappreciated colleague in Holland, Pierre Lyonet, were better observers than Réaumur himself. Réaumur set the standards by which the younger observers measured themselves, but their observations of the reproduction of aphids and the discovery of regeneration went far beyond those of their mentor. Following the demanding observational techniques that Réaumur required, both Trembley and Bonnet made discoveries concerning asexual reproduction which called into question one of the very rules which Réaumur had thought incontestable: that for animals to reproduce, mating was necessary. Throughout his career as a naturalist, Réaumur did not look for exceptions to nature's rational plan, but its underlying regularity. For this reason, in attempting to explain the apparent absence of mating, Réaumur initially suggested in his *Histoire des insectes* that perhaps aphids coupled in the womb, prior to giving birth. Not only had Réaumur failed to distinguish the male aphid, but when it was proved that aphids were truly parthenogenetic, at first Réaumur resisted the idea that aphids could produce eggs, as well as living young. Animals were supposed to be either oviparous or viviparous. Later, however, because it was proved that aphids could be both, Réaumur was determined to establish the presence of eggs in the polyp.

In contrast to the study of aphids which Réaumur clearly inspired, his role in the discovery of the polyp and Trembley's subsequent study of regeneration and budding was indirect. Without Réaumur, it is possible that Trembley's work might have passed into obscurity. However, Réaumur was slow to accept the conclusions that Trembley drew from his observations. He did not communicate the discovery immediately to the Paris Academy, but waited until he had verified it by his own study of the live specimens that Trembley was finally successful in sending from Sorgvliet. The polyp immediately created a sensation. Madame Geoffrin, hostess of one of the popular Parisian salons, wrote to Martin Folkes that it "greatly occupied Paris for a time, and above all the ignorant more than the learned for MR de Réaumur claimed that he had known about it for a long time."[2] A jar of polyps that Voltaire kept on his mantle undoubtedly stimulated many lively exchanges.

Réaumur initially challenged Trembley's description of the phenomenon of budding and urged him to consider whether he had not failed to observe a concealed egg. Consistent with Malebranche's theory of preexistence, Réaumur believed that all life came from eggs. Eggs contained one within the other were the vehicle by which life was handed down through the centuries after Creation. Nevertheless, Trembley did not rely on Réaumur's authority and persisted in his exploration of the structure of the polyp. Clearly, the perplexing reproductive phenomena that he, Lyonet, and Bonnet had encountered in their study of aphids motivated him to continue his study. He carefully described the process of budding, the connection of mother and child by a continuous membrane, and their final separation. Because Réaumur was not disposed to accept unique examples in nature, he did not give up his search for eggs and his tenacity was rewarded by the discovery that polyps did indeed produce eggs as a means of protecting the species through the winter. Thus, like aphids, polyps could be both viviparous and oviparous.

Though Réaumur was correct in maintaining that polyps produce eggs, he made a mistake in his description of the plumed polyp *(Lophopus)* that revealed his fundamental misunderstanding of the structure of the polyp. Réaumur

---

[2]Harcourt Brown, "Madame Goeffrin and Martin Folkes: Six New Letters," *Modern Language Quarterly* 1 (1940): 226.

thought that the plumed polyp produced tubes, which he called *polypiers,* piled one upon the other within which the polyps resided. He thought that these constructions were similar to birds' nests or the cells of wasps and bees. Réaumur made the same mistake in his description of corals. Réaumur thought that the soft body of the coral resided within the structures it produced, rather than being part of the living animal. Such structures demonstrated the skill or *génie* of insects, proof that they were not mere automata, but acted with limited intelligence. Whether Réaumur had difficulty grasping the implications of the discovery of regeneration for the doctrine of animal soul, or merely chose to avoid rashly committing himself, is not clear. Bonnet drove home the issue in more than one letter, but Réaumur did not choose to respond until the publication of the introduction to the last volume of his *Histoire des insectes.* In this discussion he used the very words that Bonnet had used in his letters to present the problem to his readers, but he provided little solace to those looking for answers to questions which Réaumur pointed out were metaphysical rather than scientific.

This volume was unexpectedly Réaumur's last publication which concerned insects, though he had already done considerably more research. Two more volumes were published by twentieth-century investigators still impressed with the quality of his powers of observation.[3] Why Réaumur ceased to pursue research on insects remains a matter of conjecture, though it is possible that the unsolved metaphysical issues may have determined him to give up the study of insects. Whether his research on digestion in birds and his work on the artificial incubation of chicken eggs can be related to the problems that the polyp raised for the old-style natural history practiced by Réaumur has not been investigated by recent historians of science. Indeed, although Réaumur was one of the seminal individuals for the practice of French natural history prior to the publication of Buffon's famous (or infamous) *Histoire Naturelle* in 1749, his work has been left relatively unexplored.

[3]Volume Seven of Réaumur's *Histoire des insectes* was published in 1929 in *Encyclopédie entomologique,* ed. Maurice Caullery, ser. A. XXXIIa. Morton Wheeler translated and edited Réaumur's *The Natural History of Ants* (New York: Knopf, 1926).

Though Réaumur attempted to hide his dissatisfaction with the turn toward more speculative natural history by Buffon and his "clique," his association with Abbé Joseph-Adrien Lelarge de Lignac's anonymous attack on Buffon in *Lettres à un Amériquain sur l'histoire naturelle, générale et particulière de monsieur de Buffon* (1751) shows that he was intensely interested in combatting the new materialistic theories that were emerging as a result of the discovery of the polyp. Buffon's and John Turberville Needham's championship of spontaneous generation could only have seemed to him the undoing of his life's work. Kircher's theory had returned to haunt his old age.

The letters which Trembley and Bonnet exchanged with each other are particularly illuminating because they precede the appearance of the first published works of each: Trembley's *Mémoires, pour servir à l'histoire d'un genre de polypes d'eau douce à bras en forme de cornes* (1744) and Bonnet's *Traité d'insectologie* (1745). They were written as their lifelong friendship developed and as they hesitatingly launched their careers. These letters are completely different in both tone and point of view from those addressed to Réaumur. Though Réaumur's prediction that other animals would be found that shared the strange prerogative of regeneration from cuttings influenced Bonnet's subsequent study of worms, he did not suggest specific experiments. Bonnet's experiments complemented Trembley's because he was deeply involved in the philosophical issues that Trembley's discovery had raised. Any suggestion in Trembley's letters was passionately discussed by Bonnet and his teachers, and Bonnet used hints in Trembley's letters to fashion new experiments.

Trembley's association with the empirical Newtonians of the Leiden school can be seen in Trembley's use of Boerhaave's definition of animals as inverted plants and his recognition of the difficulty of finding characteristics to clearly distinguish animal from plant. Trembley viewed the experiment of turning the polyp inside out as one of the most important that he performed and Bonnet initially seized upon it as proof that the polyp was the "point of passage" between the animal and vegetable realms. Later he would accept Trembley's conclusion that the polyp was a simple animal. The Leiden influence on Geneva was important even before Trembley's journey to Holland. Cramer had personally forged the early links with the Dutch scientific tradition when he visited 'sGravesande in 1727. Calandrini and Cramer had continued this association through their involvement with 'sGravesande's *Journal Littéraire*.

However, it is unlikely that the appeal of Boerhaave and 'sGravesande was exclusively scientific. Like Great Britain, Geneva shared with Holland and England a common Protestant background and it is certain that the scientific teachings of the Leiden Newtonians were important to the Genevan community because they represented an acceptable natural theology. The influence of natural theology on Trembley and Bonnet can be seen in their common admiration for the philosophy of Samuel Clarke. Both Trembley and Bonnet were attracted to a variety of other writers who encouraged the study of natural history because it revealed the action of God's providence. Like the Newtonian arguments for gravity as evidence of God's continuing active presence in His Creation, the study of natural history provided an antidote to the Cartesian "fiction" of a completely mechanical universe. Deeply religious, Trembley, possibly influenced by Locke, thought that God had providentially endowed matter with certain properties, like gravity, sensitivity in plants, and instinct in animals. Both Trembley and Bonnet were intrigued by the spontaneity of the behavior of their insects, and their ability to respond to stimuli such as touch and light. Bonnet initially took pleasure in his insects because their behavior appeared to indicate that they had a soul.

After Trembley's discovery, it was the possibility that the Cartesian soulless automaton would appear to be the best explanation for the ability of an animal to regenerate that prompted Bonnet's anguished reaction. Trembley's experimental hydras and Bonnet's worm with two tails demonstrated that living matter from one animal could be grafted on to another. The philosophical implications were sobering. Bonnet reported to Trembley that their professors were confounded by the enigma that the polyp presented. Not only was the doctrine of animal soul called into question, but also the view that God had designed a rationally ordered world.

The experimental creation of new individuals seemed a defiance of that order. They could not have failed to realize that others would regard their experiments as an indication that matter itself had organizing principles heretofore reserved for the Creator.

The association of Cramer and Calandrini with Louis Bourguet, an important Swiss popularizer of Leibnizian ideas, particularly the idea of a chain of being, and Bonnet's dinner with Bourguet at a crucial point in the metaphysical debate over the polyp is significant. Though Leibnizian philosophical works were difficult to obtain and their complicated metaphysics did not appeal to a generation primarily concerned with observation and experiment, this association strengthens the view that suddenly because of the philosophical problems that the polyp raised, the metaphysics of Leibniz, already known if not previously given great intellectual currency, took on new importance in Geneva. The Leibnizian view of matter as dynamic, permeated throughout with life, did away with the logical difficulties encountered by those who accepted the doctrine of animal soul within a Cartesian framework. The intermediate, sensitive substance of Pardies and Régnault could be accommodated to the Leibnizian view of a graded series of souls in nature. Rather than a mechanical *emboîtement* of the germ within the egg, Bonnet suggested in his letters that the perplexing phenomena that his worms presented could be explained by germs of heads and tails arranged in line within the body, capable of taking on the role of animal soul if stimulated by sectioning. This idea, Bonnet explained in letters to both Trembley and Cramer, came from Calandrini. How similar Calandrini's germs were to Leibnizian monads is a question which cannot be answered. Nevertheless, because of the metaphysical debate in Geneva, Bonnet was well prepared for his reading of *La Théodicée* several years later.

Trembley's polyp made the decade of the 1740s one of crisis. As Bazin, one of Réaumur's disciples, wrote in 1745 in a little work, *Lettre d'Eugène à Clarice au sujet des animaux appelés polypes, que l'on fait multiplier et produire leurs semblables en les coupant par morceaux:* "A miserable insect has just shown itself to the world and has changed what up to now we have

believed to be the immutable order of nature. The philosophers have been frightened, a poet told us that death itself
has grown pale . . ."[4] Regeneration led to the metaphysical
impasse anticipated by Bonnet's letters. If animals such as
the polyp and the earthworm were granted multiple souls
within their bodies, held in reserve until they were activated
by sectioning, why not grant an organizing principle to matter
itself? Why was a soul necessary at all? The materialist La
Mettrie could point to the polyp with satisfaction as evidence
of the soulless automaton. From the polyp to the *L'Homme
machine* was but a short step.

By the end of the century the idea of animal soul, such a
serious issue for Bonnet at the time of discovery of the polyp,
had become a subject to be derided. Thus, in 1788 Eugène
Melchoir-Louis Patrin wrote satirically:

How lucky is the earthworm, to whom are allotted. . . an infinite
number of souls, just as one more trimmed sail is added to a ship;
meanwhile we poor star-gazers must sourly drag along with only
one solitary soul! The good Descartes never dreamed that in our
century it should develop that the animals he classed as automata,
and to whom he assigned a soul no greater than a wooden wall-
clock, should thus crowd out the human race.[5]

Trembley and Bonnet were both dismayed by the turn
toward speculation, irreligion and materialism that the discovery of the polyp encouraged. Following the example of
Réaumur, and true to Newton's famous dictum, *Hypotheses
non fingo,* observation of nature, not system building, had
been the basis of their early work on insects. They had been
careful to avoid framing hypotheses. As Bonnet had written
to Trembley when considering the perplexing phenomena
of the reproduction of aphids, "Let us amass as many facts

---

[4][Gilles Auguste Bazin], *Lettre d'Eugène à Clarice* (Strasbourg: Imprimerie du Roy
et de Monseigneur le Cardinal de Rohan, 1745). Cited in Jean Torlais, *Réaumur:
un esprit encyclopédique en dehors de l'Encyclopédie* (Paris: Albert Blanchard, repr. 1961),
167.

[5]E.L.M. Patrin, *Zweifel gegen die Entwicklungstheorie,* tr. Georg Forster (Gottingen:
Dieterich, 1788), 97–119. Cited and translated by Charles W. Bodemer, "Regeneration and the Decline of Preformationism," *Bulletin of the History of Medicine* 38
(1964): 29.

as possible. There will be time enough to imagine."[6] And Trembley in the same tone, had responded to Bonnet's discovery of the perplexing worm with two tails, "It seems to me that you do well not to amuse yourself in giving reasons for the singular fact to which these worms have presented you. In general, you risk losing time in searching for explanations, and make only hypotheses, which serve for little or nothing."[7] Obviously aware of the speculative interpretation to which his enigmatic creature might be subject, Trembley firmly resisted the coupling of the polyp with materialism. As he wrote at the end of his *Mémoires,* "Nature should be explained by nature, not by our own views."[8]

By the end of the decade, less competent observers did not hesitate to let the imagination fill in the gaps in the factual record. With the publication of Buffon's *Histoire Naturelle,* it was clear that the taste for hypotheses had returned. Trembley wrote sadly to his former employer, Count William Bentinck, *a propos* of Buffon's ideas on reproduction,

M[R] de Buffon claims to explain nearly everything about generation, but I admit that I can only consider his system as a dangerous hypothesis. He tries to prove too much with the facts on which it is built. He seems to let himself be carried away by his imagination. If his work is very popular, I am afraid that he will do harm to Natural History by bringing back the taste for hypotheses.[9]

Trembley did not have long to wait for Diderot's imagined human polyps in *Le Rêve De d'Alembert.* The discovery of the polyp created a revolution in eighteenth-century ideas about the generation of animals. The repercussions of Trembley's "miserable insect" extended well beyond science into social and moral realms of eighteenth-century life. However, Trembley and Bonnet were not revolutionaries but conserv-

---

[6]Bonnet to Trembley, 24 March 1741, George Trembley Archives, Toronto, Ontario.

[7]Trembley to Bonnet, 4 October 1743, Ms Bonnet 24, Bibliothèque Publique et Universitaire de Genève. (See below, 228.)

[8]Trembley, *Mémoires,* quoted above, 134.

[9]"MR de Buffon pretend presque tout expliquer sur la generation: mais, j'avouë que je ne puis considerer son systeme, que comme une hypothese hasardée. Il fait trop prouver aux faits sur les quels il l'a batit. Il semble quelques fois qu'il se laissa emporter à son imagination. Si son ouvrage est fort gouté, je crains qu'il ne fasse tort à l'Hist. Nat. en ramenant le gout des Hypotheses." Trembley to William Bentinck, 9/20 January 1750, Ms Egerton 1726, The British Library, London.

ative Genevans who supported Calvinism and the patrician
order. They spent the rest of their lives wrestling with the
implications of their early observations of insects.

Appendix A

The Bonnet-Trembley Correspondence

Letters from Bonnet to Trembley are preserved in the George Trembley Archives, Toronto, Ontario. There are ninety-six letters written between 1740 and 1784, the year of Trembley's death. Nine of these letters are undated. Of the many letters discovered through my research, these are both the most important and the most interesting. Bonnet's letters to Trembley written between 1748 and 1761 are unfortunately missing, though the possibility exists that some of the nine undated letters are from this period. These letters were not available to Jean Torlais, Raymond Savioz or Jacques Marx when they did their research. Their full examination, therefore, promises to shed light on Bonnet's intellectual development. They were studied by Maurice Trembley with a view to publishing both sides of the correspondence, since copies of many of the Trembley letters preserved in the Archives of the Public Library of Geneva can also be found in the George Trembley Archives. These were made by Maurice Trembley's secretary, Louise Plan. Maurice Trembley published only two excerpts of Bonnet's letters to Trembley in the notes of his *Correspondance inédite entre Réaumur et Abraham Trembley:* parts of the letters of 24 March 1741 and 1 September 1741 on pp. 66 and 88. George Trembley and I look forward to collaborating on an edition of the entire correspondence.

The early letters, written between 1740 and 1744, that are published here focus on parthenogenesis and the discovery of the polyp. The Bonnet letters are published by permission of Professor Trembley. Photocopies of the original letters in the Public and University Library of Geneva from Trembley to Bonnet, written between 1740 and 1747, were graciously furnished to me by M. Philippe Monnier. These letters are catalogued as Ms Bonnet 24 with the exception of the letter of 22 December 1741 which I found in the *Dossier ouvert,*

189

autographe Trembley. I also discovered one letter from Bonnet to Trembley, 31 January 1744, in the Collection Condet, Ms Suppl. 363, fol. 8–9.

Paragraphs of the letters of 27 January 1741 and 5 May 1741 from Trembley to Bonnet are published in Bonnet's *Mémoires autobiographiques,* pp. 63 and 65. In addition, fragments of five letters from Trembley are published in the notes of Maurice Trembley's *Correspondance inédite,* located on page 21 (letter of 27 January 1741); 190 (letter of 24 March 1744); 389 (letter of 18 November 1752); 390 (letter of 8 March 1755); 392 (letter of 16 August 1755).

Trembley's correspondence with Charles Bonnet appears to have begun in 1740. The first letter (or letters) is missing. The earliest letter which has been preserved was written by Trembley from Sorgvliet, 26 July 1740. This letter appears to be a response to an earlier, now missing, letter from Bonnet. Bonnet's response to this letter is also missing. Trembley's second letter to Bonnet was written 4 October 1740 from Sorgvliet. Thereafter, with the exception of a letter, possibly written 7 August 1744, the letters written between 1740 and 1744 appear to be reasonably complete.

Though my present research comprises only the first years of the Bonnet-Trembley correspondence, the entire correspondence is of great historical interest. It fills in some of the gaps in the Bonnet-Réaumur correspondence. For example, the uncomplimentary comments about Buffon that Bonnet may have made in his letters to Réaumur, presumably destroyed by Buffon after the Réaumur legacy to the Academy of Sciences fell into his hands, can also be found in Bonnet's letters to Trembley.

I have attempted to transcribe the letters without making any changes in the original spelling or punctuation, taking Otto Sonntag's *The Correspondence between Albrecht von Haller and Charles Bonnet* (Bern: Hans Huber, 1983) as my model. Particularly bothersome was the inconsistent use of accents and capitalization. Nevertheless, I have refrained from making changes in the text, except in a few cases when the meaning was unclear without the acute accent in the final syllable of a word. I have also added an apostrophe, when necessary to improve readability. The case of *qu'elle* vs *quélle* is more delicate, as there usually appears to be a break, and the mark—whether intended as an apostrophe or an accent—is so carelessly placed as to be open to either interpretation. Under these circumstances, it seemed best to print *quélle*, which is possible eighteenth-century usage, when there was not a clear break. Because I have translated long passages from the letters in my text, I have annotated the letters very sparingly.

List of Letters from Charles Bonnet to Abraham Trembley in
George Trembley Archives

18 December 1740	20 November 1775
24 March 1741	8 January 1776
1 September 1741	13 June 1777
19 February 1741	1 July 1777
28 June 1742	4 August 1777
21 November 1742	19 July 1778
27 December 1742	18 August 1778
26 April 1743	13 December 1778
30 August 1743	30 December 1778
5 September 1743	12 February 1779
8 November 1743	30 November 1779
26 June 1744	12 December 1779
29 September 1747	23 December 1779
16 January 1748	11 January 1780
20 July 1761	28 February 1780
28 July 1761	3 November 1780
8 August 1761	23 August 1781
31 August 1761	24 June 1782 (?)
11 July 1762	3 July 1782
23 May 1763	6 July 1782
4 June 1764	22 July 1782
4 September 1764	29 July 1782
12 September 1764	10 August 1782
25 September 1764	18 November 1782
1 October 1764	22 November 1782
9 October 1764	1 March 1783

Note: Bonnet used prerevolutionary measurements. There were 12 *lignes* in 1 *pouce*. A *pouce* equaled approximately 27.5 mm. (slightly over a inch); a *ligne*, about 2.3 mm.

Tuesday 13th, 1768	29 July 1783
24 June 1768 (?)	22 August 1783
22 September 1768 (?)	5 September 1783
3 October 1768	20 November 1783
13 October 1768	February 1784 (?)
21 October 1768	6 March 1784
19 November 1768	12 March 1784
15 July 1769	16 March 1784
8 May 1772	23 March 1784
25 January 1774	25 March 1784
4 February 1774 (?)	30 March 1784
2 March 1774	6 April 1784
27 October 1774	16 April 1784
6 November 1774	26 April 1784
7 November 1774 (?)	16 May 1784
13 November 1774	
3 August 1775	
8 September 1775	
21 September 1775 (?)	9 letters which were
26 October 1775	not dated

1. Trembley to Bonnet.   Sorgvliet le 26ᵉ Juillet 1740

Monsieur

Ce seroit manquer essentiellement de gout pour l'Histoire
des Insectes, que de négliger de me procurer la rélation de
vos découvertes. Si j'en avois faites qui méritassent vôtre attention, je vous proposerois une échange; mais je n'étudie
encore qu'en écolier, et je n'ai d'autre motif à vous présenter
que vôtre propre complaisance. Je me suis surtout attaché à
suivre Reaumur et j'ai bien eu occasion de l'admirer.

J'ai de deux espèces d'insectes mangeurs d'étoffes, differens de la Teigne, l'un chenille dont il resulte un assès joli
Papillon, et l'autre ver dont il resulte un Scarabé. Si Reaumur
ne les connoit pas, il faut qu'il n'y en ait point en France. Je
suis actuellement dans une campagne, situé au milieu des
Dunes d'Hollande, qui peut être me fourniront quelques insectes particuliers. J'assure de mes respects très humbles
Monsieur Bonnet et je suis avec un parfait attachement
    Monsieur

Vôtre très humble et très obéissant
Serviteur
A. Trembley

2. Trembley to Bonnet.

Sorgvliet le 4ᵉ 8ᵇʳᵉ 1740

Monsieur

La découverte que j'ai fait de vous me fait plus de plaisir que celle d'un des insectes les plus rares. L'histoire de vôtre Hermite[1] m'a fait un sensible plaisir, et me prouve votre Sagacité et vôtre patience. Sans doute que vous réitérerez l'expérience. Un de mes amis,[2] grand observateur, l'a faite ici, cet Eté, sur un Puceron du Rosier: elle a reüssi tout comme à vous. Il y a déjà deux ans que le mémoire de Mᵣ de Reaumur m'avoit donné envie de la faire. Des expériences que j'ai faites avec quelques insectes pendant le grand froid que nous avons eu, m'ont mis en train d'étudier l'état des differens insectes pendant l'hyver. Le grand nombre de Pucerons que [nous avons] eu cet Eté et leur délicatesse aparente, m'a fait ré-soudre à chercher comment ils passent cette Saison. Je com-mence a en rassembler actuellement. Quelques uns ont ac-couché en ma présence et j'ai été tenté de mettre les nouveaux nés dans la Solitude. J'envoie à Mᵣ de Reaumur, ma chenille mangeuse de laines. Elle ressemble assés à la Teigne, elle est un peu plus grosse: de la 1ᵉʳ Classe. et les jambes membra-neuses terminées par une couronne complete de crochets. Elle n'est pas teigne. Elle est mineuse d'Etoffes. Quand elle en trouve d'assés épaisses elle les mine comme tant de vers minent les feuilles. Sinon elle joint deux morceaux d'étoffes et se place entre deux. Le papillon à la tête blanche, garnie d'un plumet blanc fort joli. Le Corselet et une tiers des ailes est brun; le reste est blanc argenté. Il est du 3ᵉ genre des nocturnes; du genre de port d'ailes en queue de Coq. On le trouve dès le mois de Juin, jusqu'à celui de 7ᵇʳᵉ. Pardon mon Cher Cousin, si je ne vous envoie que ce petit barbouillage. Je suis un peu pressé. Faites moi le plaisir de me mander vos

---

[1]This is a reference to Bonnet's experiments on parthenogenesis in which he put his female aphids "in solitude" to observe successive virgin births.

[2]Undoubtedly a reference to Pierre Lyonet who was also engaged in a study of parthenogenesis.

découvertes: cela m'évertuera à en faire. Je suis avec un veritable attachement
   Monsieur

                        Votre très humble et très obéissant
                        Serviteur
                           A. Trembley

   3. Bonnet to Trembley.   A Geneve ce 18. X^{bre} 1740

   Comme le tems des Vacances, Monsieur et Cher Cousin,
est un tems que je consacre presque tout entier aux Observations, j'ai cru que vous ne trouveriés pas mauvais que je
differat [sic] à vous repondre, qu'elles fussent expirées. Mais
ayant été obligé depuis d'écrire à Monsr de Reaumur pour
le remercier de même que l'Académie des Lettres de Correspondance dont elle m'a honoré, et pour lui faire part de
nouvelles Observations, j'ai Souhaîté d'attendre réponse afin
de vous mander son jugement sur vos Observations dont je
n'avois pas manqué de lui parler dans ma lettre. Voici donc
ce qu'il m'a fait l'honneur de me repondre là dessus. "M.
Trembley a eu la politesse de m'envoyer la Chenille mineuse
d'étoffes et ses Observations. Cette Chenille a été gravée dans
le 3e. vol. des Mem. sur les Insectes. Je l'ai rangée parmi les
fausses teignes, et j'ai parlé des ravages quêlle avoit fait dans
une de mes Brelines [sic, for berlines] dont le drap étoit fort
à son goût." Vous voyés mon cher Cousin, que vôtre attention
n'a point été indifferente à Monsr de R. et je ne doute pas
non plus qu'il n'ait été très satisfait de vos recherches. Je
m'attends qu'il vous aura écrit une lettre des plus obligeantes.
Je vous prie de vouloir bien me la communiquer, personne
n'y prenant plus de part que je n'y prends. Je vous deman-
derai encore la vôtre comme quelque chose de très curieux
pour tout amateur d'Histoire naturelle, Si je ne craignois
d'abuser peut etre de vôtre Complaisance. Quoique j'aye été
un peu Surpris de retrouver dans votre petite chenille la
fausse teigne de Monsr de Reaumur, je ne laisse pas d'étre
charmé qu'elle vous ait procuré l'occasion de lier correspondance avec ce grand Observateur et de lui faire connoître qu'il
a trouvé en vous un élève, et un élève digne de lui. Vous
voilà dès à présent engagé à lui rendre compte de vos re-

cherches et de vos observations. Je m'interesserai toujours de la maniere la plus particuliere aux éloges qu'elles vous attireront de Sa part, et j'espere que je ne tarderai pas à vous écrire sous le titre de Correspondant de l'Academie. Ne prenés point ceci, mon cher Cousin, comme une flaterie, vous n'ignorés pas combien il est avantageux pour cette Savante Compagnie d'avoir beaucoup de correspondans; c'est le veritable moyen de faire du progrès à la Physique: aussi ne neglige-t-elle aucune des occasions de s'en procurer. Les academiciens y sont même tenus par le reglement de 1699.

Je me rejouis d'aprendre ce que vous aurés découvert sur la maniere dont les Pucerons se garantissent du froid. C'est ce qui n'a point été encore bien éclairci. Je serois pourtant fort disposé à penser avec M.r de la Hire qu'ils se retirent alors dans les crevassses des arbres, ou peut être dans des trous sous terre. Après cela, il n'y auroit pas tant de quoi nous surprendre quand ils passeroient l'Hyver tout delicats qu'ils paroissent sans etre bien defendus, puisque nous savons que certaines chenilles quoiqu'encore fort petites (les *Communes*) peuvent resister à un froid de 4. à 5. degrés plus violent que celui de 1709.

Je n'ai pu qu'etre bien aise d'aprendre que l'ami dont vous m'avés fait l'honneur de me parler a fait sur les Pucerons du Rosier les memes observations que j'ai faites sur ceux du Fusain. Elles me sont une confirmation d'un fait que je ne saurois trop souhaîter de confirmer. Je vous felicite d'avoir un tel ami. Comme j'ai conçu pour lui une veritable estime, puis-je me flatter qu'il voudra bien me donner une part dans l'honneur de la sienne? Il me semble que les Observateurs n'étant pas encore fort communs, le peu qui s'en trouve aujourdhuy devroit se connoître, et se lier par un commerce reciproque pour l'avantage de l'Histoire naturelle. De mon côté je ne négligerai rien pour faire en sorte que nôtre correspondance puisse egalement nous amuser et nous instruire. Je vais donc commencer à fournir mon contingent.

J'ai continué pendant les vacances dernieres mes observations sur les Pucerons. Je me suis principalement attaché à une espece qui vit sur le chêne que j'ai tout lieu de croire du même genre que celle dont parle M. de R. T. 3. p. 334.[1]

---

[1]Réaumur, *Mémoires des insectes* 3, 334.

L'attention et l'assiduité avec lesqu'elles j'ai taché de l'observer
m'ont valu des observations et des decouvertes qui me paroissent dignes d'etre connuës. J'ai trouvé des mâles à cette
espece de Pucerons, et si petits à proportion des femelles qu'il
en est presque de même que parmi les Gallinsectes. Ils ont
4. aîles perpendiculaires au plan de position; les femelles en
sont pour la plupart depourvuës. Mais au lieu que chés les
abeilles on ne trouve qu'une seule Mere, chés nos gros Pucerons du Chêne, on trouve au contraire plusieurs pucerones
pour un seul Puceron. Ce mâle est peut etre un des plus
ardens qu'il y ait dans la Nature. Il ne semble quasi faire
autre chose que s'accoupler dès que le jour est venu. Les
femelles offrent aussi la Singularité de faire tantôt des pucerons vivans et tantôt des foetus qu'elles arrangent les uns à
côté des autres comme les Papillons disposent leurs oeufs.

Je suis actuellement extremement occupé à mettre au net
les details de ces observations que M. de R. m'a demandé
pour les communiquer à l'academie, et c'est ce qui fait que
je n'ai pas le tems de vous en écrire plus au long, non plus
que sur d'autres observations, aiant d'ailleurs d'autres occupations indispensables. M. de Reaumur qui veut bien
m'honorer d'un exemplaire de ses* d'y voir les Abeilles resoudre Physiquement un probleme que les plus grands Geomet[res] de l'ant[iquité] n'auroient pu resoudre par le secours
de leur geomet[rie]. Je dois vous dire que M. le Prof. Cramer
le resout cependant par la simple geom[etrie], ce que j'ai écrit
à M. de R. qui demande la Solution.

Je suis avec le plus vif attachement tout à vous

Bonnet

4. Trembley to Bonnet.   La Haie le 27. Janvier 1741

Vos nouvelles observations, Monsieur, m'ont fait un sensible plaisir. Il est vrai que Monsieur de Reaumur m'a écrit
une lettre très polie et au de là de ce que je mérite; surtout
ne lui aiant envoié que l'histoire d'un animal qu'il connoissoit
déja. Lorsque je trouvai cette Chenille mineuse d'étoffes il y
avoit plus d'une année que j'avois lu le mémoire de mons[r]

---

*Memoires m'envoya le mois de 7bre dernier le 5e vol. Vous devez avoir été bien
surpris

de R. sur les fausses Teignes. Plein de l'idée qu'il en devoit parler dans le mémoire où il recherche les moiens de conserver les Etoffes de laine, je ne cherchai que dans celui là. Si vous ne m'avies dit qu'il connoissoit encore un Scarabé et non une Chenille, je n'aurois jamais pensé à la lui envoyer.

Ce que vous me dites sur les foetus que font les Pucerons, nous est aussi connu ici. Monsr Lionnet, l'ami dont je vous ai parlé, m'en à montré au mois d'Octobre d'un Puceron du Chardon. Il en a encore eu d'une des espèces de Pucerons du Rosier. Cette même en [a] aussi fait dans un de mes poudriers au mois de Novembre.

La plupart de mes Pucerons sont peris, mais de faim et non de froid. J'en ai à la Campagne sur le Houx et sur le Rosier que je visite de tems en tems. La plupart ont été emportés par le vent, ou tués par la pluie. C'est dans mon cabinet que les autres sont morts de faim. J'ai nourri depuis leur naissance dans une parfaite Solitude deux Pucerons du Sureau. Ils m'ont fait des petits pendant le mois de 9bre et une partie de X[bre]. Les intervalles de leurs accouchemens ont été fort inégaux et quelques uns de plusieurs jours. Ils ont dépendu de la variation de la température de l'air. Des que le thermométre de Prins descendoit au 42 dégré ils cessoient d'accoucher, et dès qu'il remontoit, ils recommençoient à accoucher. J'ai tué par accident un de ces deux pucerons. Je l'ai ouvert, et je lui ai trouvé dans le corps plusieurs foetus qui posés sur un verre bien clair et regardés contre le jour avec une loupe, ressembloient fort à ces foetus que m'ont fait les pucerons du Rosier.

Je me suis tant soit peu émancipé à chercher la manière de la genération des Pucerons. La prémière idée que j'ai suivie est celle qu'indique Mr de R. J'ai pensé que suposé que les Pucerons s'acouplassent dans le sein de la mére, les deux qui devoient s'accoupler ne devoient pas diferer beaucoup au dégré de perfection et que si ils ne s'acouplent qu'après être parvenus à la perfection requise pour naitre, la naissance de l'un ne devoit, peut être, pas éloigner de celle de l'autre. J'ai donc été attentif pour voir s'ils en naitroit toujours deux à peu de distance: et comme les accouplemens ne sont pas si prochains dans cette Saison, elle m'a paru plus favorable pour faire l'expérience. J'ai trouvé beaucoup de variété dans

la distance des accouchemens. J'ai eu cependant a trois re-
prises deux pucerons qui se sont suivis beaucoup plus près
que les autres. Enfin je n'ai rien pu conclure et je me suis
trouvé aussi savant qu'au commencement.

J'ai formé depuis le mois de novembre le dessein d'élever
plusieurs generations de suite de pucerons Solitaires, pour
voir s'ils feroient toûjours également des petits. Dans des cas
si éloignés des circonstances ordinaires, il est permis de tout
tenter. Je me disois, qui sait si un acouplement ne sert point
à plusieurs générations? J'ai envoié à une verrerie des mo-
dèles de verres propres à renfermer des pucerons Solitaires;
et d'autres pour les réunir ensuite; pour voir s'il n'y auroit
point d'acouplemens au bout de quelques generations. Mons.ʳ
Lionnet sur ces entrefaites, m'a apris qu'aiant reunis plusieurs
Pucerons du Rosier, qui étoient chacun la troisiéme generation élevée dans la Solitude, il en a vu d'accouplès. Le male
étoit ailé et plus petit que la femelle. Quoique je compte
beaucoup sur les observations de Mons.ʳ Lionnet je crois qu'il
les faut réiterer et avec de très grandes précautions avant
que de se rien persuader. Nous comptons nous y mettre des
le commencement du Printems; et nous vous invitons à faire
les mêmes expériences de vôtre côté. Le Sujet vous apartient?
Je ne sai presque si je dois apeller plante ou animal l'objet
qui m'occupe le plus à présent. Je l'étudie depuis le mois de
Juin. Il m'a fourni des characteres assés marqués de plante
et d'animal. C'est un petit Etre aquatique. Dès que l'on le voit
pour la prémiére on s'écrie que c'est une petite plante. Mais,
si c'est une plante, elle est sensitive et ambulante, et si c'est
un animal il peut venir de bouture comme plusieurs plantes.
J'en ai coupé en trois parties. Il est revenu à chacune ce qui
lui manquoit pour etre telle que cet Etre avant que d'etre
partagé, et chacune à marché, et fait jusqu'ici tous les mouvemens que j'ai vu faire à l'animal complet. J'en ai dit quelque
chose à M.ʳ de R. et depuis que je lui ai écrit, j'ai multiplié
mes experiences et un peu poussé celles qui tendent à découvrir sa Structure. Je vous félicite et je félicite l'Académie
des Sciences de ce que vous êtes devenu son correspondant.
Je reçois actuellement une lettre de Mons.ʳ de Reaumur que
je vais lire. Il me demande de lui envoier quelques uns de
mes Etres aquatiques. C'est ce que je ferai dès que la gelée

aura cessé. Je n'en ai pas encore une histoire assés complette pour vous faire part de ce que j'ai découvert. Adieu mon Cher Cousin. Je suis tout à vous.

5. Bonnet to Trembley. à Geneve ce 24ᵉ Mars 1741

Je reponds un peu tard, Monsieur et cher Cousin, à la lettre que vous m'avés fait l'honneur de m'ecrire; mais j'espere que vous voudriés bien agréer les excuses que mon Cousin le Ministre s'est chargé de vous faire là dessus de ma part. Ma cause ne pouvoit etre entre meilleures mains. Soiés persuadé je vous prie, que si j'étois plus maître de mon tems que je ne le suis, mon empressement à vous repondre repondroit au plaisir que j'ai à m'entretenir avec vous. Des Etudes de Droit, d'Histoire Naturelle et des Observations ne laissent guères de momens vuides, et la Correspondance avec Monsʳ de Reaumur me prend plus de tems qu'elle n'en prendroit peut-étre à un autre parce que je ne sai pas faire mes lettres courtes. Aussi faut-il avoir autant de patience et de complaisance qu'en a Monsʳ de Reaumur pour ne s'en pas lasser.

Vôtre petit Etre aquatique est quelque chose de si singulier et de si surprenant qu'il me semble qu'on doit le regarder comme une des plus grandes merveilles que l'Etude de l'Histoire Naturelle puisse offrir. On peut dire que vous avés decouvert le point de passage du vegetal à l'animal. Je ne doute pas que Monsʳ de Reaumur n'ait communiqué cette observation à l'Academie et qu'elle ne vous en marque son contentement en vous offrant le tître de son Correspondant. Le grand Swammerdam, sur les traces duquel vous marchés, n'auroit assurément pas laissé tomber une pareille decouverte, lui pour qui les moindres choses étoient des merveilles. Je suis cependant bien aise qu'elle lui ait échappé, et qu'elle eut été reservée pour un Elêve de Monsʳ de Reaumur. Mais j'aurois souhaîté que vous m'eussiés dit un mot de la figure de l'animal en question, si c'en est un, de sa grandeur, de sa couleur, de l'endroit ou on le trouve et d'autres choses de cette nature qui peuvent aider à le faire reconnoître; au lieu que ce que vous m'en rapportés est presqu'une énigme, et une enigme que je ne saurois déchiffrer. Mais je m'en console aisément quand je vois ici d'habiles gens, des Savans même

comme nos Professeurs à qui j'ai été bien aise de faire voir
vôtre Lettre, y étre arretés. Elle a extremement piqué la cu-
riosité de M.ʳ De Brosses[1] Conseiller au Parlement de Dijon
qui m'avoit prié en partant d'ici, de lui écrire ce que vous
m'en aprendriés. C'est une bonne connoissance que nous
avons faite; M.ʳ de Brosses est un homme qui a des connoissances très universelles, et qui de plus est Naturaliste et Observateur. Il est fort connu de Mons.ʳ de Reaumur. Il m'a
parlé de quelques observations qu'il avoit faites sur les vers
spermatiques; elles sont curieuses. Il a observé entr'autres,
que lorsque quelques uns de ces vers se trouvent dans une
goutte de semence trop petite pour leur permettre d'y nager
comme ils auroient fait dans une plus grande étenduë, ils
avoient recours pour prolonger leur vie à un expediant assés
singulier, c'est de tourner sur eux mêmes avec une vîtesse
étonnante.

J'attends, au reste, avec impatience de savoir ce que M.ʳ de
Reaumur pense de vôtre Etre aquatique, l'ayant prié de vouloir bien me l'apprendre. Vous aurés pû sans doute lui en
envoyer un échantillon. Je lui ai aussi envoyé en dernier lieu
un assés gros paquet rempli en partie d'Insectes. Ce sont pour
la plupart des Chenilles.

Je travaille actuellement à mettre au net mes dernieres
Observations sur les Pucerons. M.ʳ de R. sur le precis que je
lui en donnai il y a quelque tems a voulu en avoir les details
pour en faire part à l'Academie. Vous pouvés compter que
je n'oublierai pas les vôtres, ni celles de Mons.ʳ Lionnet que
je suis charmé de connoître. Il me paroît assés naturel que
les accouchemens des Pucerons suivent la temperature de
l'air; mais il ne me paroit pas de même qu'ils puissent s'accoupler dans le ventre de la mere. Cette conjecture ne s'ac-
corde pas avec l'état des Pucerons renfermés dans la matrice
où ils sont non seulement baignés d'une liqueur qui ne leur
permettroit pas de s'unir, mais encore où ils sont envelopés
d'une membrane qui tient toute[s] leurs parties mieux em-
maillottées que ne le sont celles des Crisalides. A moins qu'on
ne voulut que l'accouplement des Pucerons se fit comme
Swammerdam l'avoit imaginé de celui des Abeilles, ou de

[1]Charles de Brosses, 1709–1777.

quelqu'autre maniere analogue, je ne vois pas qu'on put concevoir qu'il s'executa dans le Sein de la Mere. Et tout cela souffre mille difficultés. Je ne suis donc pas pour cette idée, quoique de M: de R., parceque je suis persuadé que la maniere dont il l'a proposée prouve assés qu'il ne la regarde point comme vraie. A l'égard de la vôtre elle me plait beaucoup et merite assurément d'etre suivie. Amassons toûjours bien des faits, il sera ensuite assés tems d'imaginer. C'est aussi ce que je me propose de faire sur nos Pucerons dès que je le pourrai, c'est à dire, dès le moi d'avril ou de may. Je leur dois trop pour ne pas me porter à ces recherches, quoiqu'un peu fatiguantes, avec plaisir. Nous nous rendrons compte mutuellement de nos decouvertes. Au reste, vous ne m'auriès pas fait chagrin de m'envoyer le dessein des verres propres à renfermer des Pucerons Solitaires et à les reunir ensuite. Ces petites inventions font quelquefois autant d'honneur à l'Observateur que les Observations elles mêmes, puisqu'elles en sont comme l'instrument. Mais c'est apparemment à cause de cela que vous m'en avés parlé si en abregé, ou plutôt que vous ne m'en avés rien dit. Je vous prierai donc de vouloir bien m'en faire part.

Vous avés vû par le Memoire de Monsieur de Reaumur sur les differentes parties des Chenilles que ce grand Observateur n'est pas du Sentiment de Malpighy par raport à l'usage des Stigmates. Celui-ci avoit pensé qu'ils servoient également à *l'Inspiration* et à *lexpiration* et M: de R. assûre qu'ils ne servent qu'à *l'inspiration*. Vous connoissés les experiences auxqu'elles il a eu recours, elles paroissent tout à fait contraires à Malpighy; je vous dirai neantmoins qu'ayant eu des occasions de les repêter sur les Chenilles mêmes que M: de R. propose comme les meilleures pour ces sortes d'experiences, elles m'ont reussi tout differemment et d'une maniere à favoriser beaucoup le Sentiment de l'Observateur Italien. Vous en jugerès par celles-ci. Ayant plongé dans un poudrier plein d'eau une de ces Chenilles nommées *Sphinx* parvenue à son dernier accroissement, j'observai que sitôt qu'elle y eut été plongée elle s'y agita, et qu'alors il sortit de grosses bulles d'air des stigmates. Ils avoient même une si grande disposition à s'ouvrir que pour peu que la Chenille se donnat de mouvement je voyais jaillir de ses ouvertures

des bulles plusieurs fois aussi grosses que la tête d'une épin-
gle. J'en voyois de même sortir quoique l'Insecte ne fut pas
dans l'eau en mettant un peu de Salive sur un ou plusieurs
de ses Stigmates. Je plongeai pour la 2de fois ma Chenille
dans l'eau et je l'y laissai pendant plus de 5. heures. Je fus
très attentif à remarquer d'où sortiroient les bulles d'air. Il
en sortit de plusieurs endroits, comme de la bouche, du dessus et du dessous du Corps, de la jonction des anneaux, et
enfin de fort grosses des stigmates, surtout des 2 premiers
ou anterieurs. Ces grosses bulles étoient comme chassées au
dehors, aussi gagnoient-elles promptement la surface de l'eau
où elles disparaissoient; au lieu que les autres restoient at-
tachées à l'Insecte. J'observai alors plusieurs fois une bulle
d'air prête à se detacher d'un des Stigmates anterieurs qui y
rentroit au point de disparoître entierement. La Chenille
l'inspiroit et *l'expiroit.* A la vérité, ce n'etoit comme je l'ai deja
dit que dans le tems où la Chenille se donnoit du mouvement
que les Stigmates laissoient sortir l'air, mais il n'étoit pas nécessaire pour cet effet de beaucoup d'agitation, le moindre
mouvement, à peine sensible, y suffisoit. Ce que Mr de R. a
observé par raport à la respiration des Crisalides, se trouve
donc ici avec les mêmes circonstances. Il n'est donc pas encore
bien prouvé que la respiration de ces Insectes s'execute tout
differemment dans l'etat de Chenille et dans celui de Crisalide. Peutétre n'est-ce que les Chenilles dont la peau est forte
et épaisse, qui laissent sortir l'air par leurs Stigmates. J'en
eprouverai de differentes especes, et j'espere que vous
m'aiderés à éclaircir mieux un article si important. Adieu
mon Cher Cousin, ecrivés-moi; tout à vous.

Bonnet.

6. Trembley to Bonnet.   La Haie le 5e Mai 1741

Vous aurez, sans doute, Monsieur et Cher Cousin, reçeu
le petit détail que j'ai envoié à Mon Frère, sur la figure de
mon animal aquatique et sur les moiens de le trouver. Je
l'apelle animal, parcequ'il est actuellement décidé que c'en
est un. C'est l'avis de Monsr de R. à qui j'en ai fait parvenir
de bien portans. Il lui a donné le nom de Polipe. J'en ai trouvé
encore une espéce et de même genre que la prémiére. Celle

ci est rougeatre. Elle est connue de Mr Bernard de Jussieu, qui en a parlé il y a longtems à Mr de R. mais qui n'a pas eu l'occasion de lui découvrir les propriétez que j'ai aperçeues à la mienne. Je n'ai cette seconde espéce que depuis douze jours. J'ai tout lieu de croire qu'elle renferme toutes les singularitez que j'ai vue dans la prémiére. Elle est plus grosse, ce qui me fait un très grand plaisir. Je pourrai par son moien verifier et pousser plus facilement mes observations sur la Structure de ces Polipes. Ce sont donc des animaux et tels que si l'on en coupe un, en deux ou en trois parties, chaque partie devient un animal complet. Je m'explique. Voila la figure du Polipe.

Ces fils *a* sont ses jambes. Si l'on coupe le polipe en quelqu'endroit du corps *b* la partie qui garde les fils vit, marche, &c. fait toutes les fonctions d'un Polipe entier, et les jambes reviennent à celle qui n'en a point au bout de 6. 7. 8. 9. 10 &c. jours; et alors elle fait aussi toutes les fonctions d'un Polipe complet. C'en est même un. La même chose est arrivée quand j'en ai coupé un en trois.

Une autre Singularité que présentent ces animaux, c'est leur maniére de se multiplier. Les jeunes sortent du corps du vieux comme les branches sortent d'un tronc. L'on ne voit d'abord qu'une petite excrescence, qui croit cha[que] jour, ensuite paroissent les jambes, et au bout de plus ou moins de tems, lorsque l'animal est complet il se détache de la mére. J'en ai eu dont il en sortoit cinq à la fois. Il m'est impossible d'entrer dans un grand détail, mais je puis prouver clairement qu'en effet le jeune sort de la mére; et qu'il n'en est point, comme de ces insectes dont les oeufs ou les petits ont été attachez sur le corps.

J'ai actuellement une moitié de Polipe. Celle qui n'a pas encore les jambes; Elle est coupée depuis 10 jours, et elle a produit deux jeunes polipes depuis qu'elle est coupée. L'un est bientôt pret à se séparer.

J'ai nour[r]i des pucerons en Solitude sur des branches de

Sureau, placées dans des tubes de verre de 5 pouces de longueur et de quelques lignes de diametre. Les tubes sont
percés par les deux bouts. L'un trempe dans l'eau et l'autre
est bien bouché avec du coton. J'en ai fait faire de plus grands
et dans les quelles je pourrai insinuer une petite branche,
lors même qu'elle sera encore sur le pied. Les verres dans
les quels je compte en nourrir en Société, sont de très grands
Gobelets percés au fond, pour faire passer la branche qui
devra nour[r]ir les pucerons. Le haut sera bien fermé! Je ne
sai, si j'aurai le tems de mettre cette année, ces préparatifs
en usage: Vos doutes sur la maniere d'aspirer et d'expirer
des Chenilles sont très naturels. Les expériences que l'on fait
sur ce sujet demandent bien des précautions, parce que mille
choses peuvent nous faire illusion. L'attention avec les quelles
vous les faites, rendent inutiles les soins que d'autres pour-
roient se donner. M.ʳ de Brosses pense-t-il que ces vers spermatiques qu'il a vu tourner dans une goute, sont les germes
humains? Cela me paroit sujet à bien des difficultez.

Adieu mon Cher Monsieur. Je suis à vous de tout mon
Coeur.

Trembley.

7. Bonnet to Trembley.   A Geneve ce 1.ᵉʳ 7.ᵇʳᵉ 1741

Je vous fais mille excuses, Monsieur et cher Cousin, de ce
que je suis resté si longtems sans vous écrire. J'ai été malade,
et depuis que je suis retabli je n'ai presque eu que le tems de
communiquer à Mons.ʳ de Reaumur les details de quelques
Observations qu'il m'avoit demandé il y a près d'un an, et
que je n'avois pû encore achever de mettre au net. Elles
roulent la plupart sur une nouvelle partie que j'ai observée
dans plusieurs Chenilles de differentes classes, et qui est placée entre la lévre inferieure et la premiere paire des jambes.
C'est une espece de mamelon qui à l'ordinaire est retiré dans
l'interieur du Corps, mais qu'on oblige à se montrer en pres-
sant un peu fortement la partie anterieure de l'Insecte. Il
m'a offert des varietés très remarquables soit à legard du
nombre, de la forme, de la Structure, &c. dont je vous ferois
part si la longueur des ces details m'en laissoit le loisir. Reste
à decouvrir l'usage de cette partie; sa position et sa figure

semblent d'abord lui donner bien de l'air d'une filière, c'est aussi ce qui a paru à Mons! de Reaumur par raport à une semblable partie qu'il a observée dans les teignes aquatiques, mais je crois m'étre convaincu qu'elle n'en fait point la fonction. C'est comme vous le voyés avoir encore bien peu avancé.

Mais pour parler d'Observations tout autrement interessantes il faut parler de celles que vous avés faites sur le Polype. On peut dire avec verité que vous avés decouvert dans cet étrange animal un Monde de Merveilles, qui n'eut peut étre jamais été connu sans vous. Aussi ne sais-je ce que j'y dois admirer le plus, ou l'extreme singularité des faits, ou le talent d'observer de celui qui est parvenu à les voir. Vous voilà dès à present, au nombre de ceux qui par la beauté et la Singularité de leurs decouvertes ont merité de passer à la posterité la plus reculée. Le succès de vôtre experience sur un Polype partagé en plusieurs parties suivant sa longueur m'a étrangement surpris. Quoiqu'elle fut de celles que je m'étois proposé de tenter, je n'avois pas cru devoir esperer que l'oeconomie animale n'en scroit pas detruite. Il n'etoit en effet, guères naturel de penser qu'une telle prerogative eut été accordée à des Animaux, surtout lorsqu'on voit que parmi les Plantes il n'y a presque que quelques Racines, comme celles de Chicorée à qui elle soit propre. J'avois même regardé au commencement cette épreuve comme une de celles qui pouvoient aider à décider si l'animal en question n'etoit pas plutôt une Plante. A l'égard de la reproduction du Polype coupé transversalement, comme je n'ai pû encore parvenir à m'en procurer, je l'ai au moins verifiée sur un Insecte aussi aquatique d'un genre fort different, sur un ver long de près de 2. pouces, de couleur rougeatre melée de brun, et je suis actuellement occupé à varier ces experiences le plus qu'il m'est possible. J'aprends aussi que M! Lionnet en a fait une semblable sur un ver qui me paroit peu different du mien, s'il n'est le même. Je suis charmé de me rencontrer souvent avec cet habile Observateur. J'en ai plus de confiance en mes observations; car lorsqu'il s'agit de faits autant au dessus des regles ordinaires que le sont ceux-ci on craint toujours de n'avoir pas assés bien vû.

Les Polypes que vous élevés successivement en solitude vous éclairciront apparemment sur ce que vous devés penser de vôtre soupcon touchant l'accouplement des Pucerons. Je commence à douter qu'il soit necessaire dans plusieurs.

L'industrie que vous avés remarqué dans vôtre Polype ne fera pas sans doute grand plaisir aux Metaphysiciens; si d'un côté elle semble prouver qu'il a une ame, de l'autre sa reproduction extraordinaire fait naitre de terribles difficultés. Y auroit-il dans cet Insecte comme chés ceux qui lui ressemblent dans cette reproduction, autant d'ames qu'il y a de portions de ces mêmes Insectes qui peuvent elles mêmes devenir Insectes parfaits? Ces ames dans cet état n'ont-elles encore qu'une simple idée de leur existence; ne peuvent-elles s'acquitter de leurs fonctions que lorsque l'operation a donné lieu aux germes dans lesqu'els elles sont renfermées de se developper? C'est en gros le Systeme de Mr Calandrini. Je ne sai ce qu'il vous en semblera, mais il ne me paroit pas, non plus qu'à quelques uns de nos Messrs, s'accorder bien avec mes Observations. Sitôt après l'operation j'ai vû une des portions d'un de mes vers, celle qui avoit seulement gardé la queuë, agir comme celle qui avoit gardé la tête. J'ai fait d'autres observations dont le Succès a été le même ou à peu près.

Vôtre corps organique nous tourmente; nous vous aurions été fort obligés si vous eussiés voulu nous en parler un peu plus au long. Serois-ce encore là quelqu'animal à qui la Nature ait donné de se multiplier a la facon de certaines Plantes?

Il me tarde fort d'apprendre l'Histoire des petites Anguilles dont se nourrit vôtre Polype. Je me sens du penchant à Soupconner qu'elles peuvent appartenir de près aux vers que j'étudie et qui comme ce dernier se reproduisent après avoir été partagés. Monsr de Reaumur Soupconne aussi que les Orties et les Etoiles de mer jouissent de la même prerogative. Il faut s'attendre qu'à mesure que les Observations se multiplieront cette propriété deviendra plus commune. Mais quelque decouverte qu'on fasse par la suite sur une matiere si interessante ce sera toujours à vous qu'on en sera redevable. Adieu, mon Cher Cousin, on ne sauroit être avec une plus grande estime et un plus sincere attachement que je le suis

> Votre très humble et très obéissant serviteur
> Bonnet

8. Trembley to Bonnet.   La Haie le 22 X^bre 1741

La partie presqu'imperceptible que vous avez découvertes [sic] à plusieurs Chenilles, merite certainement, Monsieur, que vous poussiez vos recherches à cet égard. Elle est tombée en de bonnes mains. Vôtre sagacité et vôtre patience nous en rendront raison. Le bonheur que mes Polipes ont de vous plaire, n'est pas un de leur moindre mérite. Je suis bien faché que vous n'en aviez pas encore trouvé. Ils sont certainement dignes de vôtre attention, et vos observations auroient pu beaucoup avancer leur Histoire. Vous me permettrez bien de ne pas entreprendre ici le détail et les preuves de singu-laritez qu'ils renferment. Il faudroit un volume. J'en suis encore fort bien pourvu à présent. Mon cabinet n'est plein que de poudriers à Polipes. J'en ai qui sont nez dans le mois de Juin, et qui depuis peu de jours après leur naissance n'ont pas cessé jusqu'à présent de multiplier, et cela sans accou-plement. Il me paroit que la maniére dont les Polipes se multiplient est extrémement analogue à celle dont les plantes multiplient par rejetton. Ces animaux singuliers sont cer-tainement encore plus admirables que vous ne pouvez l'ima-giner. J'ai fait des hydres à 7 et a 8. Têtes, en les coupant seulement en partie suivant leur longueur à commencer par la tête. J'ai ensuite fait l'exploit d'Hercule. J'ai valeureuse-ment coupé les 7 tétes à un Hydres [sic]. Il lui en est revenu sept autres, et qui plus est chacune des sept têtes coupées sera bientot [en] état de devenir un Hydre. J'ai soin pour cela de les bien nourrir. J'ai un Polipe qui est déja coupé en 36 parties, et la plupart de ces 36 multiplient, et toutes multi-plieroient si je le voulois. J'ai nommé cet autre genre de Polipes dont je vous ai parlé des Polipes à Panache. Il mul-tiplient aussi par rejetton et d'une maniére très curie[use . . .] sur la description que j'en ai faite à Mr de Reaumur, Mr Bernard de Jussieu se les est rapellé, et lui en a d'abord fourni une grande quantité; Mr de Reaumur en a ensuite trouvé beaucoup en Poitou. Il en est charmé. La Cl[asse] des Pol[ipes] est bien nombreuse. Mr de R. vous aura peut être ecrit, que Mr de Jussieu et Mr Guetar [Guettard] en ont trouvé plusieurs Espéces sur les cotes de Normandie et de Poitou.

J'en ai encore 4 autres espéces dont trois sont certainement connues par M.ʳ de Reaumur.

Le ver que vous coupez est très curieux. Il me paroit que c'est le même que celui de M.ʳ Lyonnet. Ils sont très communs ici. Il me sont d'une grande ressource pour nourrir mes Polipes. J'en ai des milliers dans un vase dont le fond est plein de terre. Ils se tiennent ordinairement dans la terre, et en sortent quelques fois. Ordinairement une partie de leur corps sort de terre, presque perpendiculairement au niveau, et elle se donne toutes sortes de mouvemens d'inflexions. Je vois qu'ils se multiplient comme mes petites anguilles; c'est a dire d'une maniére très singulière. Voici en deux mots comme celles ci le font. A quelque distance de leur bout postérieur, il se forme une tête, en sorte que cette partie posterieur devient une anguille et qu'elle paroit être attachée par sa tête à la queue d'une autre; ensuite elle se sépare &c.

Adieu mon Cher Cousin Je suis tout à vous.

Trembley

9. Bonnet to Trembley.   A Geneve le 19ᵉ Fevrier 1742

Des occupations indispensables dont je ne suis pas encore debarassé et l'état de ma santé m'ont empeché, Monsieur et Cher Cousin, de repondre plutôt à la lettre que vous m'avés fait l'honneur de m'écrire le mois de Xᵇʳᵉ dernier. J'ai lu avec admiration vos nouvelles Observations sur les Polypes et nos MM.ʳˢ à qui j'en ai fait part ne les ont pas moins admirées. L'experience d'en couper suivant leur longueur m'avoit toûjours parû essentielle: Vous l'avés poussée aussi loin qu'il falloit pour la rendre incroyable. L'Hydre des Poëtes n'est plus rien en comparaison des vôtres, et le Heros qui delivra le monde ne meritoit pas suivant moi, à plus juste tître, d'etre celebré, que l'Observateur à qui nous devons tant de prodiges. Dès que les especes de ces animaux admirables sont si nombreuses l'esperance d'en trouver me revient. Je ne m'y épargnerai pas.

Je ne saurois encore vous dire bien précisement si mes Vers sont les mêmes que ceux de M.ʳ Lionnet. M.ʳ de Reaumur à qui j'en ai envoyé et qui m'en avoit demandé m'aidera peut ètre à le savoir. Permettés moi de renvoyer à une autre occasion de vous en parler en detail. Je serai sans doute plus en état alors de vous dire quelquechose de precis.

Vous voulés savoir des nouvelles de mes observations sur les Pucerons. Je n'ai pas fait à cet égard l'année derniere tout ce que j'aurois souhaité. J'ai seulement confirmé ma 1ere Experience et commencé à en tenter d'autres. J'ai elevé successivement dans une parfaite Solitude jusqu'à la 4e generation des Pucerons du Sureau. Tous ont engendré. Les blés en nos Cantons ont été attaqués l'automne derniere d'une sorte de maladie assés singulière et qui me paroit avoir du rapport avec ce que Mr de R. a rapporté du Gramen T. 4. p. 382 de ses Mem[oires]. Sur toute la longueur des feuilles ou à peu près se remarquoient des petites taches oblongues de couleur jaune, qui observées à la loupe ne sembloient être qu'un amas d'une poussiere jaune extremement fine et semblable à celle des Etamines. Ces poussieres paroissoient sortir de l'interieur des feuilles. Les tiges ne m'en ont jamais fait voir. Je les ai beaucoup observées au Microscope et avec une attention particuliere. Elles m'y ont paru moins regulieres que celles des Etamines; mais jamais je ne suis parvenu à y decouvrir d'Insectes, ni de ces oeufs en forme de bateau dont parle Mr de R. dans l'endroit cy dessus. Vers le milieu de 7bre et encore vers la fin il falloit chercher pour trouver les feuilles attaquées de cette espece de maladie, mais dans le courant d'8bre et de 9 ^{bre} elle avoit tellement gagné que les blés de verts en étoient devenus jaunes. On distinguoit cependant des endroits où il avoit conservé sa couleur naturelle, comme au dessous des grands arbres, ainsi que j'ai cru le remarquer. En X^{bre} la poussiere avoit disparu, mais à la place des petites taches jaunes on Voyait comme de petites cicatrices qui sembloient avoir été faites par l'eruption de cette poussiere. Je ne sais point encore ce que je dois penser là dessus. Peut ètre en aurois-je été instruit, si j'avois pû continuer les Observations que j'avois commencées. Est-ce une maladie de la Plante? Ou est-ce un effet qui doit ètre attribué à des Insectes? Les Paysans accusent pour la plupart les Rosées froides. Pour moi je ne vois pas comment on peut l'entendre.

Mr de R. vous aura sans doute appris que c'est très réel que les Etoiles et Orties de mer se multiplient ou peuvent ètre multipliées de boutures quelque grands que soient ces

Insectes. Il est inutile que je vous dise que j'ai trouvé que les
Vers de terre peuvent l'etre aussi. Ce sont des experiences
que vous ou M.ʳ Lionnet avés déja faites. Mais ce qui est très
sur c'est qu'on ne sauroit ètre avec un attachement plus vif
T[out] a V[ous].

C.B.

10. Bonnet to Trembley.   Tonnex le 28 Juin 1742

Le beau present que vous allés faire au Public, Monsieur
et Cher Cousin, en lui donnant l'Histoire de vos Polypes; et
qu'elles obligations n'auront pas les Curieux à M. de Reaumur d'avoir contribué par Ses Sollicitations à la leur procurer!
Aussûrément si jamais ouvrage aura remporté tous les Suf-
frages ce sera celui-ci. Quélle foule de merveilles ne va-t-il
pas offrir! Et que le merveilleux gagnera, accompagné des
graces du Stile et de celles de la nouveauté! Je dois ajouter
l'élégance et la bonté des Figures. Ordinairement les Observateurs ne rencontrent que des dessinateurs; c'est beaucoup
même lorsqu'ils les rencontrent exacts. Vous avés dans M.
Lionnet et un bon dessinateur et un grand Observateur. Rien
ne scauroit donc vous manquer pour rendre l'Histoire des
Polypes un Chef d'oeuvre en son genre. Je ne perds point
patience à en chercher, quoique mes soins n'aient pas encore
été suivis de succès. Je me propose de faire de nouvelles
recherches dès que mes Etudes et ma santé qui n'est pas des
plus robustes me le permettront. Après cela si je ne suis pas
plus favorisé de la Fortune j'aurai recours à M. de R. Paris
étant plus près d'ici que la Haie.

Par contre mes Vers aquatiques ne cessent point de me
fournir ample matiere à barbouiller bien du Papier. J'ai envoyé en diverses fois à M. de R. les details de mes Observations sur ces Insectes, avec un Exemplaire d'Iceux. Il les a
trouvé, comme je m'en doutois, d'une espèce differente de
ceux qu'il a le plus suivis. J'ai aussi partagé de ces derniers,
au moins ais-je partagé des Vers que je crois être les mêmes,
mais qui m'ont paru se reproduire beaucoup plus lentement
que les miens. Sans doute parcequ'ils sont plus gros. Je ferois
un petit Volume, Mon cher Cousin, si je vous faisois le recit
entier de ce que j'ai vû. Je me bornerai aux resultats.

1°. J'ai observé avec attention soit à la vuë simple, soit avec le secours des Verres, la structure de ces Vers. Quoiqu'en apparence des plus simples elle offre des particularités dignes d'être remarquées. Celles qui ont rapport à la digestion et à la circulation sont des plus curieuses.

2°. J'en ai coupé en 2. en 3. en 4. en 8. en 10. parties et toutes sont devenuës autant d'animaux parfaits. Comme ils sont fort effilés, il n'est guères possible de les partager suivant leur longueur, comme vous l'avés fait avec tant de succès sur les Polypes. J'ai ensuite essayé de pousser la division plus loin et jusqu'en 24. et en 26. parties, mais comme je ne l'ai encore pratiqué qu'en Hyver j'attends de l'avoir tenté aussi en Eté pour étre mieux en état de decider sur le succès. Plusieurs neantmoins ont repris, mais elles sont ensuite peries. D'autres, ce qu'on jugera apparament très remarquable, ont vecu plus de 3. mois sans s'ètre complettées et par consequent sans avoir pû prendre de nourriture. Qu'elles ressources n'a point la Nature pour conserver la vie à de telles portions sans le secours d'alimens étranges! J'ai vû le sang circuler dans de semblables portions, quoiqu'à peine longues d'une ligne, et cela avec autant de regularité que dans l'animal entier.

3°. La tête reparoit ordinairement avant la queuë; elle cesse de croître lorsqu'elle a atteint la longueur de 1. lig.[ne] à 1. lig.[ne] 1/2.

4°. Celle-ci s'etend journellement mais les plus grands progrès se font dans les premiers mois qui suivent celui où s'est fait l'operation. Des Portions de ces Vers qui en Juillet passé n'avoient qu'environ 2. lig.[nes] ont aujourdhuy plus de 2. pouces. J'ai essayé de faire sur ces Vers ce que M. Hales[1] a fait sur les Sarmens de Vigne, je veux dire de dresser l'Echelle de leurs accroissemen[ts]. J'ai envoyé à M. de R. une Table où se trouve celle des Portions de 4. semblables Vers partagé l'un en 2., l'autre en 4., le 3e en 8. le 4e en 10. Vous comprenés assés mon cher Cousin, que je ne pretends pas qu'il soit aussi aisé d'avoir des mesures actuelles de l'extension de ces Insectes qu'il l'est d'en avoir de celle des Plantes. Trop de difficultés—et de difficultés presqu'insurmontables, s'y opposent. Je me suis contenté d'une exactitude Physique. J'ai

---

[1]In the margin of the letter is written: Statique des Veget[aux].

dit comment je m'y suis pris, les précautions dont j'ai fait usage. C'est aux Observateurs à juger après cela du degré de justesse. Par cette Table il paroît entr'autres, qu'on n'observe pas de difference bien considerable entre les progrès des moitiés et ceux des quarts, des huitiemes, des dixiemes; que pendant l'Hyver la reproduction se fait la moitié ou environ plus lentement qu'en Eté; que la derniere portion, celle de la queuë, est ordinairement celle qui toutes choses d'ailleurs égales fait le moins de progrès; qu'il faut un long tems pour que les parties nouvellement reproduites acquierent la couleur de l'ancienne.

5°. Ce n'est pas une règle que les portions qui ont repris tête, parviennent toûjours à reprendre une queuë. Ce n'en est pas une non plus que la 1ère portion, celle qui garde la tête, fasse constamment le plus de progrès. Ce qu'il y a seulement de certain, c'est que l'etat du Ver, le nombre des divisions et quelques autres circonstances, occasionent bien des variétés, dont j'obmets le détail.

6°. En partageant le mois de Juillet passé un de ces Vers en 8. parties, j'ai vû sortir d'une des portions une espece de filet extremement menu, que j'ai reconnus pour un petit ver tout semblable quant à l'essentiel au grand. Je ne desesperai pas de l'élever. J'y ai réussy. Je l'ai suivi pendant 2. mois au bout desquels un accident imprévû me l'enleva. Il avoit cru environ du double. Ces vers sont donc Vivipares. En observant avec une grande attention leur intérieur, j'ai crû distinguer des petits vers pareils à celui que j'avois fait venir au jour. J'ai cru les voir s'agiter en divers sens; mais certaines parties que j'ai remarquées et qui leur ressemblent me tiennent justement en Suspens.

7°. Des tiers de mes vers, devenus eux mêmes de grands Vers m'en ont donné de petits pareils à celui dont je viens de parler. J'ai étudié leur structure. J'y ai decouvert quelques singularités. J'en ai partagé, ils se sont reproduits avec beaucoup de promptitude. J'ai [. . . ?] fait ensuite l'experience de n'en partager qu'à demi, de façon que les portions ne tenoient l'une à l'autre que par un fil. En une heure de tems environ tout s'etoit rejoint de maniere qu'on ne voyait plus qu'une [. . . ?]. J'ai fait aussi la même experience sur les grands.

8°. J'ai commencé à faire une experience qui peut eclaircir

bien des questions. J'ai coupé constamment à un même ver la tête et la queuë à mesure qu'elles se sont reproduites. Le Corps n'a point encore cessé de se reproduire. J'en suis dejà à la 5ᵉᵐᵉ coupe.

9°. J'ai cherché s'il n'y a point dans ces Vers quelque point où si on les coupe la reproduction n'a pas lieu. J'en ai trouvé deux, à 1. ligne env[iron] des 2. extremités. J'ai donc taché d'en rendre raison, ou pour mieux dire, j'en ai indiqué une. Enfin, mon Cher Cousin, j'irois loin si je vous disois tout ce que j'ai vû et ce qu'il me reste à voir sur ces Vers et sur d'autres. Il n'y a pas du mal que nous ne nous communiquions pas nos observations plus en détail, la verité en brillera davantage. Souvent on cherche plus à voir ce que d'autres Observateurs ont vû que ce qu'on devroit voir. Dès qu'on a été prevenu il arrive qu'en voulant observer ce qui a été dejà observé on neglige d'autres observations interessantes. En un mot, on gagne toujours à interroger la Nature, de plusieurs façons, les reponses en fournissent plus de lumières. Au reste, je suis charmé, mon cher Cousin, de m'être rencontré avec vous par raport aux Observations sur les Pucerons. Celles que vous indiqués m'ont toujours parû demander d'être faites. Elles doivent procurer des eclaircissemens sur plusieurs points de leur Histoire. J'ai actuellement en Solitude la 5ᵉ generation de Pucerons du Fusain. Je pousserai aussi loin que faire se pourra. Je suis avec un entier devoument, Monsieur et Cher Cousin, t[out] a v[ous].

Bonnet.

Pardonnés je vous prie mon griffonage, écrivant fort à la hâte.

11. Trembley to Bonnet.   à Sorgvliet le 23 aoust 1742

Il y a longtems, mon Cher Cousin, que je n'ai entendu parler de vous. Ce billet est destiné à vous reveiller. Monsʳ Lyonet a fait une découverte qui vous interesse sur ces gros pucerons de chene que vous avez beaucoup observés, et parmi les quels vous avez vu des mâles en automne.

Nous nous promenions ensemble le mois d'avril dernier dans les bois de Sorgvliet, et Mʳ Lyonnet qui voit tout, découvrit sur l'écorce d'un chesne, de petits corps oblongs et brunatres, qui avoient beaucoup l'air de ce que vous prenez pour des foetus avortez de pucerons, et qu'il prend pour des oeufs. Il jugea d'abord que ces corps en étoient, et les porta dans son cabinet, d'où en effet il a vu sortir des pucerons, qui se trouvent ètre de cette grande espèce décrite dans le 3ᵉ vol. de Mʳ de Reaumur.

Ces Pucerons se sont fort multipliés sur un Chesne ici, sur le quel il y avoit des oeufs. Mʳ Lyonet les visite de tems en tems. Ils ne font point d'oeufs à présent mais des petits; et Mʳ Lyonet ne désespére pas de les voir pondre cet automne après les avoir vû accoucher pendent l'Eté!

On grave à force mes Planches. Je voudrois déja qu'elles fussent faites, car je commence à trouver que mon entreprise traine un peu.

Adieu mon Cher Cousin. Je suis tout à vous.

Trembley

12. Trembley to Bonnet.   a Sorgvliet le 14ᵉ 7ᵉ 1742

Les observations que vous avez faites, Monsieur et Cher Cousin, sur le Ver que vous avez coupé sont extrèmèment curieuses. Je vous suis très obligé de m'en avoir fait part. Elles ont à divers égards beaucoup de rapport à celles de Monsʳ Lyonnet. J'ai surtout fait une grande attention à ce que vous dites de la multiplication de ces vers, parcequ'elle me paroit différente de celle de ceux de Monsʳ Lyonnet, et de deux autres espéces, que je connois. Les nôtres se séparent d'eux mêmes. La partie postérieure de leur corps devient un ver, et quand il est bien formé il se sépare. J'ai observé cela avec soin dans une espéce surtout, dont la tête est très remarquable, et j'ai vu cette tête se former dans un jeune; c'est à dire, que j'ai vu un anneau du Ver Mére, situé à quelque distance du bout postérieur, devenir une tête; j'ai suivi ses progrès, et enfin le ver s'est séparé.

Je souhaite très fort que vous trouviés des Polypes. Vous avez d'autant plus raison de ne vous pas rebuter en en cherchant, qu'il arrive souvent que dans un fossé où il y en a eu beaucoup, l'on est quelque tems sans en pouvoir trouver un seul. Il m'a été impossible d'en retrouver de la prémiére espéce que j'ai connue: de celle avec la quelle j'ai commencé mes expériences.

Je serois surpris que l'on n'en trouvat point dans vos quartiers, car c'est un animal fort commun. Il a été vu en 1703 à Delft par Lewenhoek [sic] et en angleterre par un anonime. Ils ont ape[rçu] sa manière de se multiplier. Si vous avez les transa[c]tions Philosophiques vous trouverez deux articles sur ce sujet dans l'année 1703. No. 283 art. 1 ct No. 288. ai t. 1. Je n'ai pas encore parcouru la suite des Transactions Philosophiques, en sorte que j'ignore si elles renferment de nouvelles observations.

Adieu mon cher Cousin. Je suis tout à vous.

A. Trem[bley]

13. Bonnet to Trembley.   A Geneve le 21. 9ᵇʳᵉ 1742

Je vous felicite, Monsieur et Cher Cousin, et j'en felicite en même tems la Patrie de l'Election de mon Cousin l'avocat pour Auditeur. Quoique cette Charge ne pût guères lui manquer, c'est toûjours un Sujet de joye dans les termes où en sont encore les Choses; et un bon augure pour l'avenir.

J'ai été véritablement surpris, je vous l'avoüe d'apprendre par la lettre que vous vous étes donné la peine de m'écrire que vos Polypes étoient dejà connus dès 1703. Vous n'en avés cependant pas moins la gloire de l'invention. Et puis de la multiplication *par rejetons* à celle de *boûtures* il y a loin. Il vous étoit reservé de nous decouvrir tant de merveilles.

J'attends avec impatience l'Histoire de ces Insectes si singuliers. Avance-t-elle? Je vous remercie de m'avoir appris que dans des fossés où il y avoit beaucoup de ces Polypes, on vient ensuite à n'y en plus voir. Cela me donne de nouvelles esperances. M. de Reaumur m'écrit qu'il n'y a rien de si commun que ce genre d'Insectes sur les Côtes du Poictou. Il y en a de 2. sortes. Les uns qui sont doués de la faculté *locomotive,* les autres qui en sont privés. Ceux-ci qui ressemblent encore mieux à des Plantes ont été pris pour tels par les Botanistes. Nous pouvons les leur revendiquer aujourdhuy.

Je continuë à pousser mes Experiences et Observations sur mes vers. J'en ai decouvert de nouvelles Especes qui m'ont offert des faits nouveaux. Un des plus singuliers est celui qui

m'a été fourni par quelques portions d'une espece de ces
Insectes, qui au lieu de reprendre une Tête ont repris une
Queuë. En telle sorte que j'ai eû, et que j'ai encore des Vers
à deux queuës. Vous êtes peut étre tenté de soupconner ici
quelque illusion, quelque ressemblance assés imparfaite, et
telle qu'on en voit quelquefois dans plusieurs productions
soit du Regne animal, soit du Vegetal. Point du tout. Cette
queuë qui a poussé à la place de la tête, est absolument telle
sur toutes ses parties que l'est la veritable queuë de ces Vers.
Je m'en suis assuré par plus d'une observation. C'est je vous
assure quelque chose de plaisant que de voir l'embarras de
ces pauvres vers auxquels ce malheur est arrivé. J'en ai
d'autres qui me feront peut étre voir encore la même singularité. Je ne manquerai pas de les suivre. Voilà bien de
quoi conjecturer.

J'ai envoyé en dernier lieu à M. de Reaumur un ample
detail de mes nouvelles Observations sur ces sortes de Vers,
et sur les *Pucerons*. J'ai élevé jusqu'à la 5eme Generation de
ceux du Plantain et jusqu'à la 6eme de ceux du Fusain dans
une parfaite Solitude. J'ai essayé d'en réûnir d'ailés et de non
ailés de differentes generat[ions]. Je n'ai point vû d'accouplement. J'ai découvert dans une espece que j'ai beaucoup
suivie des mâles ailés et des mâles qui sont absolument privés
d'ailes. Leur ardeur pour la propagation de l'espece est remarquable. Je n'ai pû à cause de quelques obstacles que je
ne detaillerai pas, pousser plus loin les premières Experiences
sur les generat[ions] elevées en Solitude. Mais j'ai des Pucerons d'une autre sorte que je veux tacher de conserver pendant l'Hyver. Ce sont de ceux dont j'ai parlé, et que j'ai le
plus suivi. Tout ceci est fort curieux.

J'envoye à M. de Reaumur les details de 40. Experiences
ou Observations, sur la maniere dont s'opere la Respiration
dans les Chenilles. Il y en a de curieuses, et qui me paroissent
bien prouver que les Stigmates donnent entrée et sortie à
l'air. Je voudrois pouvoir entrer plus dans le detail.

Je ne vous ai peut étre pas dit qu'il y a déjà quelques années
que j'avois medité le plan d'un Essay sur les Insectes qui
renfermeroit en abregé ce qu'on a decouvert de plus interessant et mes Observations. M. de Reaumur à qui je l'ai communiqué m'a enhardi à Suivre ce Plan; mais je crains d'en-

treprendre un ouvrage au dela de mes forces. J'y donnerois une methode de diviser des Insectes que je puis dire en partie nouvelle. C'est celle de Swammerdam rectifiée. Je l'augmenterois de quelques Classes. J'avois commencé à mettre la main à l'oeuvre. Mais jusqu'ici j'ai presque plus detruit qu'édifié. Il est difficile qu'on parvienne a se contenter lorsqu'on veut ne rien faire que de bon. Et aujourdhuy on y regarde de bien près. Je suis, Monsieur et Cher Cousin, avec toute l'estime et l'attachement possibles entierement à vous.

Charles Bonnet.

Des 26emes de mes Vers qui reviennent de boutures m'ont donné de ces petits vers dont je vous ai parlé dans ma précédente Lettre. Je soupconne qu'ils peuvent aussi se multiplier par rejettons. J'ai eû des 26e à qui il poussoit au dessus de la tête comme un petit ver qui est ensuite rentré. Je vous en dirai davantage dans une autre Lettre si je vois quelque chose de nouveau.

14. Trembley to Bonnet.   à La Haie le 11e X^{bre} 1742

Le détail de que vous avés eu la bonté de me donner, Monsieur et Cher Cousin, de vos observations, sur les vers aquatiques que vous avez coupés, et sur les Pucerons, m'a fait un véritable plaisir. Je vois que vous ne vous relachez point, et que vous nous apprendrez de belles choses. Vôtre ver à deux queues est admirable, mais il ne m'étonne pas, parce que rien ne m'étonne. Monsr de Reaumur à vu des vers de Mr Lionnet; ils sont de même genre que les vôtres, mais d'une espéce différente, et plus grands qu'aucuns de ceux qu'a vû Mr de Reaumur, jusqu'à présent. Il m'a fait part des découvertes qu'il a faites sur les Polypes, sur les côtes de Poitou, et de celles de Mr de Jussieu sur les côtes de Normandie, qui sont très curieuses. Quoique très à portée de la mer, je n'y en ai pas encore été chercher. J'ai crain, que ce que je trouverois ne fit diversion aux Soins que je dois donner aux animaux que je suis à présent.

J'ai fait encore plusieurs expériences sur les Polypes, dont le détail prendroit trop de tems et de place. Leur corps est un Sac, il peut se comparer à un boiau: je l'ai retourné comme l'on retourne un bas ou un gant. L'intérieur est devenu l'ex-térie[ur] et l'extérieur, l'intérieur, et les Pol[ipes] sur les quels j'ai fait cette opération, vivent, mange[nt] et multiplient. Adieu mon Cher Cousin, je suis tout à vous. Pardon de ce que je vous écris si succinctement, je suis pressé.

A Trembley

15. Bonnet to Trembley.   A Geneve le 27e X^{bre} 1742

En verité, Monsieur et Cher Cousin, vous poussés nôtre admiration à bout. Quoi! tous les jours de nouveaux Pro-diges! Ne craignés vous point enfin qu'à force de les multi-plier on ne s'y accoutume, et qu'on ne vienne à les regarder d'un Oeil assés tranquille? Il me paroît que vous vous êtes déja bien familiarisé avec Toutes ces merveilles. Quoique je tache de m'y faire, je vous avoue qu'il y a des momens où j'en suis si frappé qu'il me semble n'avoir rien vû. Voilà donc des Animaux qui, non seulement peuvent être multipliés de boutures, et dont les Petits sortent du Corps comme une branche d'arbre sort du Tronc, mais encore qui peuvent être retournés ni plus ni moins qu'un bas et qui ainsi retournés ne laissent pas de vivre de se nourrir et de multiplier. Vous êtes bien malicieux de ne nous en pas dire davantage et de ne nous parler jamais que par Enigmes. Monsr Cramer, et tout ce qu'il y a ici de Savans et d'Honnetes Gens qui vous admirement font le même reproche. Vous n'avés point de Conscience de mettre de cette facon nôtre Esprit à la Torture. En mon particulier je Souhaîterois fort de savoir comment vous avés été conduit à faire une Semblable Experience et comment vous vous y êtes pris. Je ne comprends pas comment vous l'avés pû tenter avec Succès sur des Polypes entiers, je le concois mieux de Portions: A la verité la Structure de ces Insectes ne m'est guères connuë encore. Je desirerois aussi que vous m'aprissiés si ce retournement merveilleux arrive naturellement, comme je suis fort porté à le croire. Mes Vers aquatiques sur lesquels j'ai beaucoup operé se divisent quelquefois sans que je m'en mêle. Et il estoit bien naturel que ces Insectes n'eussent pas besoin de la main de l'Obser-vateur pour être mis en état de faire usage de telles facultés. Vos Polypes se raccourcissent extraordinairement: ne leur arrive-t-il point de se raccourcir encore davantage et de se retourner enfin? N'est-ce point aussi dans l'instant de la contraction que vous les avés pris pour faire l'experience qui nous surprend? Le Corps de ces Insectes seroit-il formé d'une double membrane dans la duplication de laquélle la Nature auroit placé les visceres et toutes les parties necessaires à la vie? Il faut suivant Mr Cramer qu'ils aient des veines lactées, ou l'équivalent, tant au dehors qu'au dedans. Enfin les Polypes retournés croîssent-ils également vite et multiplient-ils comme ceux qui n'ont point été retournés? Peuvent-ils de même être multipliés de bouture, et en tout Sens? Les parties de la Generation des mâles abeilles, le Cerveau des Limacons, les Cornes &c., les Orties de Mer, les Insectes qui passent par l'etat de *boule allongée,* &c. fournissent comme vous savés des Exemples de retournemens analogues. Mais il y a encore loin de là à vos Polypes. Quand nous en donnerés vous l'Histoire? Voila bien des Questions. Vous y repondrés lorsque vous en aurés le tems. Nous serons contens pour peu que vous veuillés être moins laconique. Adieu, Monsieur et Cher Cousin, je vous embrasse de tout mon Coeur, et suis avec l'attachement le plus Sincère entièrement à vous. C. Bonnet. Vous aurés de mes nouvelles une autre fois. Je voulois dire de celles de mes Vers.

16. Trembley to Bonnet.   à La Haie le 8e Fev. 1743

Si vous aviés des Polypes, Monsieur et Cher Cousin, je me ferois un plaisir et un devoir de vous indiquer la maniére dont je les tourne. Mais n'en aiant point, je vous prie de permettre, que je vous renvoie, et tous les Curieux de Genéve, au petit ouvrage au quel je travaille et qui paroitra peutetre cette année, si la gravure des planches ne l'arrète pas trop. Je fournis dans ce païs ci des Polypes à tous ceux qui en Souhaitent, et je leur indique tous les expediens que je mets en usage pour faire mes expériences, afin qu'ils puissent les revoir et les perfectionner. J'ai appris à retourner des Polypes à Monsr Allamand, un de mes amis, qui demeure à Leyde et qui est connu à Genéve. Il le fait aussi bien que moi.

C'est l'idée, quoi que très vague, que j'ai de la Structure des Polypes, qui m'a fait venir la pensée de les tourner. Je l'ai déja entrepris en 1741 mais sans succès. J'y suis revenue l'Eté dernier, et aprés bien des tentatives pénibles, j'ai trouvé un moien trés court et très facile à emploier. Les Polypes ne se retournent jamais d'eux mêmes. C'est sur un Polype entier que je fais l'expérience. Il paroit que l'etat où se trouve un Polype après avoir été retourné ne lui plait point. Il entreprend presque toûjours de se détourner, et souvent il réussit. J'ai ordinairement recours à un expedient pour empécher les Polypes de se détourner. J'enfile un Polype retourné à une Soie de porc, je la passe près de son bout antérieur, près de ses lévres: quand elles veulent se renverser, elles sont arrétées par la Soie de Porc. Ce n'est rien pour un Polype que d'être ainsi enfilé. Il mange et multiplie comme si de rien n'étoit, avec Sa Soie de porc qui passe au travers de son corps.

La plupart des Polypes retournés sont en état de manger quelques jours après l'opération.

Un Polype retourné ne différe en rien d'un Polype qui ne l'est pas.

Je ne suis pas surpris que vos vers se séparent d'eux mêmes. J'ai vu cela dans les mille pieds à dards, dont parle Mons.ʳ de Reaumur dans son sixiéme volume, et qu'il a fait multiplier par la Section. Il y a longtems que j'ai écrit à Mons.ʳ de Reaumur que ces mille pieds à dards se séparent naturellement, et que c'est là une de leurs maniéres naturelles de multiplier. Je ne connois que celle là.

Cette espéce de vers à des anneaux bien marqués. Si l'on en observe un de suite, l'on ne tardera pas de s'apercevoir, que peu à peu un des anneaux, situés plus prés du bout postérieur que de l'antérieur, se forme en tête; l'on voit pousser cette espece de trompe que cet animal a à son extrémité antérieure. Elle sort perpendiculairement au corps du mille pied. L'on voit aussi paroitre les yeux trés distinctement. Enfin cette tête est si parfaitement formée, qu'au lieu d'un seul mille pied à dard que l'on voioit au commencement, l'on en découvre deux. La partie postérieure de la mére est devenue un jeune, qui lui est encore réuni, et dont l'extrémité antérieure, dont la pointe de la tête, est unie à bout de la queue de l'autre mille pied. Enfin le jeune se sépare et peu de tems après, l'on voit encore la partie postérieure de la mére devenir un autre mille pied à dard, qui se sépare comme le prémier.

Vous jugés bien qu'il faut que la mére mange et croisse pour pouvoir multiplier. Il y a plusieurs espéces de vers, qui je suis sur, sont du genre des vôtres, et à qui il m'a paru, qu'il avoit été donné de se multiplier de cette manié[re]. C'est une espéce de ces vers qui me sert à nourrir mes Polypes pendant l'hyv[er] mais je ne me suis pas attaché à les observer. L'on ne les coupe pas dans mon Cabinet, mais on les dévore.

Adieu, Monsieur. Vous ne vous plaindrés pas cette fois que je suis laconique. Il est trop tôt pour vous promettre un exemplaire de l'Hist[oire] des Polypes. Je ferai en sorte que vous en aiés un presqu'aussitôt qu'elle paroitra.

17. Bonnet to Trembley. A Geneve le 26. Avril 1743

Vous aurés sans doute appris, Monsieur et Cher Cousin, par Monsieur le Ministre vôtre Frere les Raisons qui m'ont empêché de repondre plutôt à la Lettre que vous m'avés fait l'honneur de m'écrire le 8ᵉ Fevrier dernier. J'ai été extremement occupé à me mettre en état de Subir les Examens d'Avocat, au nombre desquels on a bien voulu me recevoir. Maintenant que me voila debarassé d'un fardeau qui me pesoit beaucoup, je vais retourner à nôtre Histoire Naturelle avec un nouveau plaisir. Je pense dèja à chercher des Polypes, et j'espere de réussir à en trouver. Certains endroits vers lesquels je n'ai pas encore dirigé mes recherches m'en fourniront peutétre. Après tant et de si étonnans prodiges qu'ils vous ont montré vous jugés bien mon Cher Cousin, qu'il ne m'est pas possible de rester indifferent sur leur Compte. Aussi y a-t-il longtems que le desir de faire connaîssance avec eux me tourmente. Mais je vous l'avouë, avant que j'eusses terminé mes Etudes de Jurisprudence, je craignois d'en rencontrer. Il est sur qu'ils m'auroient été un grand Obstacle à ma Reception. Il ne m'auroit pas été possible de m'empecher de leur consacrer une bonne partie de mon tems et peutétre plus, ce qui m'auroit fort reculé.

Vous faites donc l'etonnante Operation du retournement sur des Polypes entiers et pour les empêcher de se retourner vous leur passés près des Levres une Soye de porc. Vous ajoutés que ce n'est rien pour un Polype d'étre ainsi enfilé, qu'il mange et multiplie comme si de rien n'étoit. Voilà qui me confirme de plus en plus dans le Sentiment où je suis que de tous les animaux les Polypes sont de ceux dont la Structure est la plus Simple et se raproche le plus de celle des Plantes. Il me semble que si on avoit à dresser l'Echelle des Etres Organisés, on devroit les placer immediatement au dessus des Vegetaux. Mais que dans cette Simplicité ces Insectes sont étonnans, et que la Sagacité de celui qui est parvenu à nous devoiler tant de merveilles est digne d'étre admirée!

La nouvelle Espece de Vers aquatiques que j'ai découvert pourra aussi nous fournir bien des faits curieux. Je crois vous avoir dit que j'ai vû des Portions de cette sorte de Vers prendre une Queuë au lieu d'une tête. J'ai repeté l'observation déja trois fois. J'en ai un qui vit avec deux queuës depuis le mois de 7bre et qui n'a rien perdu de son agilité. Mais j'ai voulu m'assurer si en faisant la section ailleurs que là où je l'ai faite dans ceux dont je viens de vous parler je ne parviendrais point à remettre les Choses dans l'ordre, et c'est ce qui est arrivé. Il y auroit bien des Reflexions à faire sur tout ceci, mais que vous ferés, mon Cher Cousin, mieux que moi. Je ne manquerai pas de continuer à varier ces Expériences. Elles le meritent trop. Je vous informerai ensuite du resultat.

J'ai fait l'Automne derniere de nouvelles Observations sur les gros Pucerons du Chesne dont Monsieur de Reaumur a inséré les premières dans le 6e Volume. Si je ne craignois d'abuser de vôtre Complaisance mon cher Cousin, je vous prierois à cette occasion de me dire ce que vous pensés des Conjectures de Monsr de Reaumur sur ce sujet. Elles ont assurément quelque chose de singulier. Mais j'ai de la peine à les admettre. Il est pourtant vrai qu'un Puceron de cette Espece que j'ai élevé en Solitude n'a pu se defaire que d'un de ses fetus; mais je l'attribuë au froid et non au defaut de mâle. Les Experiences nous en apprendront d'avantage. J'aurois été même bientôt en état de decider si j'avois pû en conserver pendant l'Hyver comme je me le promettois. J'ai envoyé à Monsr de Reaumur avec ces Observations, une manière plus commode d'élever des Pucerons en solitude. Elle consiste à remplir à moitié un Poudrier d'eau, à appliquer sur l'ouverture un rond de Carton percé vers le milieu d'un trou par lequel on fait passer la petite branche qui doit four-

nir la nourriture au Puceron, et à en recouvrir d'un autre
Poudrier dont l'ouverture s'applique exactement sur le trou
du Carton. S'il reste entre eux quelque vide je le fais disparoître en garnissant tout le tour de sable sec. Vous comprenés
aisément mon cher Cousin l'avantage de cet expedient. Le
principal consiste en ce que le petit Puceron ne risque point
de se noyer et que s'il lui prend fantaisie de quitter la branche
il peut toujours voyager sans qu'on le perde jamais de vuë.
Il y a longtems que je m'en sers avec succès. On peut employer
pour le mieux des Cartons de differentes couleurs suivant
celle du Puceron.

Les nouvelles Experiences et Observations Sur la Respiration des Chenilles que j'ai communiquées à Monsr de Reaumur lui ont parû meriter attention. Il me fait l'honneur de
me dire *qu'elles lui ont rendu douteux ce qu'il croyoit très certain.*
Je pourrois vous en parler une autre fois un peu plus au
long. Vous vous rapellés de quoi il s'agit. Je me propose de
les repêter et Monsr de Reaumur se l'est aussi proposé. J'en
medite d'un autre genre qui, si elles reussissent, auront de
quoi nous Surprendre.

Je vous embrasse de tout mon Coeur, Monsieur et cher
Cousin et suis avec le devouement le plus entier vôtre très
humble et très obéissant Serviteur. Bonnet.

Le Thermomètre de M. de Reaumur que je suis assès as-
sidument a été le 27. X^{bre} à 7.h.1/2 Mat[in] à 6. deg[rés] et le
4e Fevrier à la même heure à 7.1/2. Les maladies ont regné
ici dès le commencement de l'Hyver et ont emporté beaucoup
de monde.

18. Trembley to Bonnet.   à Sorgvliet le 31 Mai[?] 1743

Monsieur et tres Cher Cousin

J'ai bien des remerciements à vous faire de vôtre derniére
lettre, qui m'a fait un sensible plaisir. Quelque briévement
que vous me parliez des [objets de?] vos observations, il m'est
très facile de juger, de la dextérité, de la patience, et de la
Sagacité que vous y emploiez. Mr Folkes président de la Société Roiale de Londres m'a parlé plusieurs fois dans ses
lettres des observations que vous avez envoiées à Monsr Hans
Sloane. Elles ont été fort goutées: et je n'ai pû que féliciter

très fort M.ʳ Folkes de ce que la Société recevoit des Mémoires d'un observateur tel que vous. Le fait de ce ver à deux queues est des plus singuliers, et mérite d'être suivi avec une grande attention. Je verrois avec un tres grand plaisir le détail de vos observations sur la respiration des chenilles: mais je n'ose de vous les demander. Quand penserez vous à rassembler les observations que vous avez faites et à les faire imprimer?

Je souhaite ardemment que vous trouviez des Polypes: Je suis persuadé que vous me seriez d'un grand secours si vous en aviez, pour compléter mes expériences. Il y a beaucoup d'autres sortes de Polypes d'eau douce dont M.ʳ de Reaumur vous aura peut être parlé. Ils sont fort differens de ceux qui ont fait le principal de mes recherches. Je ne parle pas seulement de ceux que j'ai nommés Polypes à Panache,[1] dont il est fait mention dans la préface de 6.ᵉ Vol de l'hist[oire] des Insect[es] mais de diverses autres especes beaucoup plus petites, qui se trouvent sur tous les corps qu'on tire des fosses. L'on tire quelque fois de l'eau des morceaux de bois qui paroissent couverts de quelque chose de blanc que l'on prendroit pour une sorte de moisissure: C'est un nombre prodigieux de Polypes, que M.ʳ de Reaumur à apellés Polypes à entonnoir. Il y en a d'autres qui forment un petit bouquet, parce que les tiges qui soutiennent leur entonnoir sont attachés les unes aux autres. M.ʳ de Reaumur connoit trois espéces de ceux à entonnoir, des blancs, des bleus et des verts; je ne connois que les blancs et les verts; j'en connois diverses especes de ceux à bouquets. Il faut les observer avec une forte loupe, pour bien distinguer leur figure. Je leur ai vu avaler des animaux. Vous jugez bien qu'ils étoient d'une grande petitesse. Je crois qu'il y auroit de belles découvertes à faire sur tous ces Polypes; c'est pourquoi je désire extrémement que vous en trouviez: persuadé qu'ils ne sauroient tomber en de meilleures mains.

Adieu mon Cher Cousin. Je suis tout à vous, de tout mon coeur.

19. Bonnet to Trembley.   A Tonnex le 30ᵉ Aoust 1743

J'ai renvoyé jusqu'ici, Monsieur et très cher Cousin, à repondre à vôtre obligeante Lettre, parce que j'attendois d'avoir

---

[1] Lophopus

quelque chose de nouveau à vous communiquer. Il est vrai que j'avois remis à Mons.ʳ le Professeur Jallabert qui me l'avoit demandé, un memoire pour Mons.ʳ le Chevalier Sloane, et qu'il a eû le bonheur de lui plaire et à la Societé Royale. Mais je suis bien éloigné de penser qu'il puisse être comparé à celui que vous avés envoyé et que j'ai eû le plaisir de lire avec Mons.ʳ le Prof.ʳ Cramer dans les *Trans. Phil.*[1] C'est, à mon gré, un Chef d'oeuvre. Je n'ai jamais rien vû de plus precis ni de plus net. Mais j'y ai encore plus admiré l'etonnante Sagacité de l'auteur. On ne peut donc que feliciter extremement la Societé Royale de Londres de ce qu'elle est prête à posseder un membre tel que vous. Je l'ai appris recemment, et je vous en felicite aussi. Elle veut bien m'accorder le même honneur, mais je le redoute presque, parceque je sens parfaitement qu'il s'en faut beaucoup que je sois en état de remplir les devoirs d'un bon Academicien. Je n'ai a lui offrir qu'assés de patience, et encore plus de bonne volonté. Lorsque vous aurés occasion d'ecrire à Mons.ʳ Folkes vous voudrés bien l'assurer de mes très humbles respects.

L'Observation du Ver à deux queuës meritoit assurement d'étre poussée. Je ne l'ai pas non plus negligée. Je l'ai même repetée plus de fois qu'il n'étoit naturel de l'attendre. Lorsque j'ai coupé de ces Vers en deux parties, la 2.ᵈᵉ ou n'a point repris, ou a repris une queuë a la place d'une tête, et il en a été de même lorsque j'en ai partagé en 3. et en 4. portions. Il n'est pas encore tems de chercher à rendre raison de ce fait Singulier; il faut pour cela avoir amassé plus d'observations et d'Experiences et c'est à quoi je travaille. S'il falloit moins de force, de vigueur dans ces Vers pour pousser une queuë que pour pousser une tête on verroit quelque jour à expliquer ce Phenomene embarassant. C'est la premiere idée qui s'est offerte à mon Esprit et qu'il me paroit qu'on peut destruire ou confirmer par une Experience que j'aurai soin de tenter.

Voici une autre Observation assés curieuse. Dans le mois de Juin dernier il me tomba entre les mains un Ver de l'espece de ceux que j'ai le plus suivi, dont la longueur étoit d'environ 2 pouces 1/2. L'ayant mis dans un vase à part avec de l'eau

---

[1]Abraham Trembley, "Observations and experiments upon the fresh-water Polypus," *Philosophical Transactions* 42 (1743): 3–11.

et un peu de terre, je le trouvai quelques jours après partagé en 3. parties. Toutes trois avoient commencé à se completter lorsque je remarquai à l'extremité anterieure de la 3ᵉ une espece de petit mamelon. J'attendis plusieurs jours pour voir s'il se developeroit et s'il ne donneroit point une seconde tête comme j'etois fort porté à le soupçonner; mais comme il demeuroit le même, je fis l'experience d'en couper le bout. Mon dessein étoit de determiner par là les sucs nourritiers à s'y porter en plus grande abondance pour faciliter le developement. Et en effet, j'eû peu de jours après le plaisir d'avoir un ver à deux têtes, toutes deux bien formées.

Vous étes allé bien autrement loin avec vos Polypes; vous en avés fait à beaucoup plus de têtes. Mais mes vers ne sont pas si traitables. Leur extreme agilité, et leur molesse ne permettent pas de tenter sur eux des Experiences semblables à celles que vous avés tentées avec tant de succès sur les Polypes. On ne peut ici qu'aider la Nature comme j'ai essayé de le faire. Je vais pousser cette Experience. Aussi je m'apercois que les volontés ne sont pas les memes dans ces deux têtes, lorsqu'une tire d'un côté l'autre tire de l'autre. J'ai actuellement en solitude la 6ᵉ generation des Pucerons du Plantain, et je suis prêt à en former la 7ᵉᵐᵉ Je tiens un Registre des accouchemens pour chaque Generation.

Je me suis enfin determiné à donner un precis de mes observations les plus interessantes. Mais je crains pour le succès, surtout venant après vous. Monsʳ le Profʳ Neker m'a prié de vous faire ses complimens, et il vous sera obligé si vous voulés bien assurer de ses respects Monsieur le Comte de Bentink. Je prends la liberté de lui presenter aussi les miens que vous aurés la bonté de lui faire agréer. J'ignorois qu'à ses autres grandes qualités ce Seigneur joignoit encore celle d'Observateur. Vous êtes bien heureux d'avoir un tel confidant des Secrets de la Nature. Il est bien difficile qu'elle échape à des yeux si clairvoyans. Je finis en vous assurant qu'on ne peut étre avec plus de devouement et d'estime que je le suis, Monsieur et très cher Cousin, entierement à vous Charles Bonnet. Point encore de Polypes.

20. Bonnet to Trembley. A Tonnex le 5. 7bre 1743

Vous verrés, Monsieur et Cher Cousin, par ma Lettre du 30. du Passé, la raison pourquoi je ne vous ai pas écrit plutôt. Celle-ci est pour vous remercier de celle par laquélle vous me donnés part de la decouverte de M. Lyonnet. Vous dites très bien quand vous dites qu'elle m'interesse. Lisés, je vous prie, depuis la page 558. du 6eme Vol. des Mem[oires]. Sur les Insectes, jusqu'à la page 560, et vous y verrés que c'est M. de Reaumur, qui regarde ces petits Corps oblongs dont accouchent quelquefois les Pucerons, comme des fetus avortés, et que pour moi j'ai toûjours eû plus de panchant à les soupçonner des oeufs. C'est donc avec un plaisir bien Sensible que je vois ma conjecture se vérifier. Je ne manquerai pas de chercher à repeter l'Observation de M. Lyonnet. J'ai même déja fait à cet égard, l'année derniere et la precedente plus d'une tentative. Mais la saison, a [sic] ce que je juge, ne m'à [sic] pas été favorable. M. Lyonnet peut compter que ces Pucerons du Chêne pondront des oeufs cette automne. Je me crois fondé à le lui prédire.

Je pense bien que le tems que vous employés à l'Histoire des Polypes ne lui fera rien perdre. Mais le Public doit s'impatienter. Je vous trouve heureux d'avoir un bon dessinateur et un bon graveur à vôtre disposition; je n'ai ni l'un ni l'autre. Et je ne sais pas trop encore comment je m'y prendrai pour faire graver les Planches qui doivent entrer dans mon Recueil. Mais ce ne sont pas les seuls Secours qui me manquent et que vous avés.

Je vous embrasse, Monsieur et Cher Cousin, et Suis tout à vous.

C. Bonnet.

21. Trembley to Bonnet.   à Sorgvliet le 4ᵉ 8ᵇʳᵉ 1743

Monsieur et très Cousin

Les derniéres observations que vous avez faites sur les vers
qui peuvent se multiplier par la section, sont extrémement
curieuses: si je n'étois pas déja trop occupé, j'aurois de la
peine à m'empécher, de couper quelques vers aquatiques,
afin d'avoir le plaisir d'observer ce [sic, for ceux] à deux
queues que vous avez suivis avec tant d'attention. Connoissez-
vous un peu particulièrement la forme et la structure de la
tête, de cette sorte de vers, et les avez vous vu manger? Ceux
qui ont deux queues ne peuvent apparemment prendre aucune nourriture. Il me paroit que [vous?] faites très bien de
ne pas vous amuser à rendre raison du fait singulier que vous
ont présenté ces vers. On risque ordinairement de perdre
son tems à chercher des explications, et on ne fait ordinairement que des hypothèses, qui ne servent à rien ou à peu de
chose. Vôtre ver à deux têtes merite certainement toute votre
attention.

J'ai appris avec beaucoup de plaisir, que vous poussiez
constamment vos observations sur les puccrons. Le sujet est
trop beau pour l'abandonner. Les Pucerons du chesne que
Mʳ Lyonnet observe, ont beaucoup multiplié cet Eté, et ils
ont toujours été vivipares. Le tems de la ponte de leurs oeufs
approche. Vous pensez bien qu'on ne négligera pas de les
observer.

Vous ne pourriez, sans vous exposer à de justes reproches,
vous tarder plus longtems, à faire imprimer vos principales
observations. Je voudrois bien pouvoir contribuer à faciliter
l'exécution de vôtre entreprise. Trouverez-vous à Genéve un
Libraire qui puisse les imprimer; et comment ferez-vous si
vous n'avez ni dessinateur ni graveur? Si vous aviez un dessinateur, je vous conseillerois de m'envoier vos desseins, et
je les ferois graver ici, de maniére que vous en seriez content.
Je sai depuis longtems que vous avez été proposé à la Société
Roiale, et je me flatte d'apprendre bientôt la nouvelle de vôtre
élection. J'ai eu l'honneur d'être élu le mois de mai dernier.

Je joins ici une petite planche, qui renferme des Polypes.
C'est l'ouvrage de Mʳ Lyonnet, qui réussit dans tout ce qu'il
entreprend. Il s'est avisé de graver, et le premier morceau

qu'il a fait s'est trouvé excellent. Il a gravé cette planche trois semaines après avoir touché un burin pour la premiére fois de sa vie. Peut être, que quand vous aurez une idée plus précise des Polypes, il vous sera plus facile d'en trouver.

La fig. A. représente un Polype suspendu à la superficie de l'eau *a* par son bout posterieur. Les fils déliés qui sortent de l'autre extrémité, lui servent de bras; c'est avec ces bras, qu'il arrache les animaux qui les viennent rencontrer. Le bras *b* a arrété un mille pied F qui se debat. Le Polype le rapproche peu à peu de la bouche *c* en contractant et en recourbant le bras *b*. Ce Polype A est à jeun.

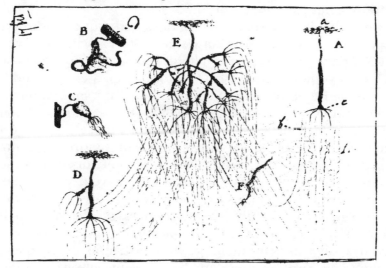

Celui de la figure B est fixé par son bout posterieur contre un brin de bois *d*. Il est en train d'avaler un ver beaucoup plus long que lui. Il l'avale en double, ce qui l'oblige à ouvrir extrément sa bouche.

La fig. C représente un Polype qui a achevé d'avaler sa proie et dont la bouche est entiérement refermée.

La fig. D fait voir un Polype dont il sort un petit, et celui de la figure E en pousse plusieurs à la fois de differens endroits de son corps. Il y a plus, quelques uns mêmes de ces petits, en produisent déja d'autres.

Je n'ai pas besoin d'entrer dans un plus grand détail sur l'explication de ces figures, parce que le petit mémoire que vous avez lû dans les Transactions Philosophiques, vous aura, j'espére, assez mis au fait, sur les principaux traits de l'histoire des Polypes.

On grave à force mes Planches, j'en ai déja reçeu quatre, dont je suis très content.

Adieu, mon cher cousin, je suis de tout mon coeur Tout à vous.

Mons.ʳ le Comte de Bentinck a été très sensible au souvenir de M.ʳ Necker. Il m'a recommendé de lui faire parvenir bien des complimens de sa part. Voulez vous bien le saluer très humblement de la mienne.

22. Bonnet to Trembley.   Tonnex le 8.ᵉ 9ᵇʳᵉ 1743

Monsieur et très Cher Cousin.

Je vous fais mille remercîmens de la Lettre que vous m'avés fait le plaisir de m'écrire, et de la petite Planche dont vous avés bien voulu l'accompagner. M.ʳ Lyonnet, à ce que je vois, n'a qu'à vouloir pour réussir en tout ce qu'il entreprend. Il est, je pense, le premier Naturaliste qui ait joint à la qualité de grand Observateur, celles d'habile Dessinateur et d'habile Graveur. Avec de si rares talens, que j'admire comme je le dois, que lui reste-t-il à Souhaîter? Sans prétendre le disputer en aucun point à M.ʳ Lyonnet, j'ai essayé toute fois, de dessiner moi-même les principales figures qui doivent entrer dans mon petit ouvrage. Mais comme je n'ai jamais appris le dessein, j'en ai envoyé un échantillon à M. de Reaumur pour Sçavoir s'il les trouvera passablement éxactes. Dès qu'il m'aura dit son sentiment, je vous instruirai de ce que je compte de faire. En attendant, Souffrés que je vous temoigne ici ma reconnoissance de l'offre obligeante que vous avés la bonté de me faire à ce sujet. L'envie de voir mes figures bien exécutées, pourra bien l'emporter chés moi sur la crainte de vous devenir incommode. J'aurois fort souhaité que mon ouvrage eut pû paroître à peu près en même tems que le vôtre. Comme nous ne nous sommes communiqué aucun détail, le Public auroit du plaisir à voir en quoi nous nous serons rencontré, et en quoi nous differerons. N'étoient les figures, le mien pourroit paroître dans les premiers mois de l'Année prochaine.

J'ai été fort occupé, et je le suis encore, à un Memoire sur le Solitaire. Il s'y agit principalement d'une decouverte qui vient d'étre faite recemment d'un Specifique très sur pour l'expulser. M. Herrenschwandt, medecin de Morat en Suisse, est le possesseur de ce beau Secret. Après avoir reussi soit dans nôtre ville, soit en d'autres endoits de la Suisse, sur plus d'une 40^{taine} de sujets, il va par l'Allemagne en Hollande et en Angleterre pour revenir par la France dans sa Patrie. Je compte lui donner une Lettre de recommandation pour vous; et j'espere, Mon Cher Cousin, que vous serés bien aise, de même que M. Lyonnet d'avoir une si belle occasion d'observer un Insecte aussi digne de vos Recherches que l'est le Solitaire. Mon Memoire, ou pour parler plus juste, ma dissertation est divisée en 3. Parties. Dans la premiere il n'est question que de la maniere dont M. Herrenschwandt administre son Remede et des diverses circonstances qui en accompagnent l'opération. La Seconde comprend mes Observations sur cet Insecte. La troisieme est la discussion des Questions Suivantes, qui me semblent étre les plus interessantes de celles qu'on peut agiter sur le Tenia. Qu'elle est son Origine? Comment se propage-t-il? Y en a-t-il de plusieurs Especes? Est-ce un seul et unique animal, ou une Chaîne de Vers? Repousse-t-il après avoir été rompu? Peut-il y en avoir plusieurs dans le même Sujet? J'ai dejà envoyé à M. de Reaumur, les deux premiers parties, avec les deux premieres Questions de la 3eme. Je vais travailler aux autres. Vous jugés bien que je ne suis pas encore assés instruit pour dissiper tous les doutes. Je me contente de faire voir où reside le plus probable. Le tems et les Experiences de M. Herrenschwandt joints aux Observations des Naturalistes et aux vôtres en particulier, nous en apprendront davantage. Il est heureux qu'il y ait à la Haye deux observateurs, comme vous et M. Lyonnet: Car je ne doute pas qu'il ne s'y trouve bien des Gens travaillés du Solitaire. Il est commun en Hollande. Comme M. Herrenschwandt doit arriver au premier jour à Strasbourg, j'en ai averti M. Bazin, et l'ai invité à observer ce ver extraordinaire. Ainsi je m'attends à étre bientôt éclairci sur beaucoup de points de son Histoire.

Adieu, mon Cher Cousin, je vous embrasse de tout mon Coeur et suis avec le plus parfait devouement, entierement à vous

Charles Bonnet.

23. Trembley to Bonnet.   à La Haie le 19ᵉ Xᵇʳᵉ 1743

Monsieur et très Cher Cousin

Si vous n'avez pas à Genéve un bon graveur, je crois qu'il vous faudra avoir recours, à un Ouvrier de ce païs-ci et vous ferez d'autant mieux, qu'il n'y en a actuellement aucun dans le reste de l'Europe, qui puisse graver des Planches d'Histoire naturelle comme ceux que nous avons ici. Mais il est pour cela absolument nécessaire qu'on leur fournisse de bons desseins, et qu'on les paie bien. J'ai une Planche dans mon ouvrage qui coute seule cinquante florins d'Hollande, et elle ne contient qu'un Polype grossi au microscope, et trois morceaux de bras de Polype, aussi grossis au microscope. Quand vous voudrez m'envoier un échantillon de vos figures, je vous dirai si on peut les graver comme il faut, et à peu près, ce que coutera la gravure. J'ai emploié un nommé Van der Schley Disciple de Picart. Il est très habile. Au reste vous pouvez compter que je ne traiterai aucun sujet dans mon ouvrage, que vous traitiez dans le vôtre. Il ne s'agira dans le mien que de Polypes. Vous voiez donc, que rien au monde ne peut vous engager à souhaiter que nos ouvrages paroissent en même tems.

Je juge par l'extrait de la dissertation sur le Solitaire, qu'elle sera très curieuse. J'ai vu ici un jeune medecin de Morat nommé Herrenschwandt, qui a étudié à Leyde, et qui a ensuite éxercé sa profession à Tournai. Seroit-ce celui là qui à fait la découverte du Secret?

Quand ce Monsieur sera ici, je me ferai un sensible plaisir de lui être bon à quelque chose: et je souhaite de tout mon coeur qu'il délivre nos patiens de leurs hôtes incommodes, et qu'il me fournisse l'occasion de les examiner.

Adieu Mon Cher Cousin, je suis de tout mon coeur Tout à vous.

A.T.

24. Bonnet to Trembley.   A Geneve le 31 Janvier 1744

Monsieur et très Cher Cousin

Je vous fais mille et mille nouveaux remercimens des offres obligeantes que vous avés la bonté de me faire au sujet de mes Planches. Je m'en prévaudrai avec d'autant plus de plaisir et d'empressement que je suis bien persuadé qu'elles ne sçauroient étre mieux éxecutées nulle part qu'en Hollande et sous vos yeux. M:̲ de Reaumur à qui j'avois envoyé un échantillon de mes Desseins les a approuvés de manière à me rassurer beaucoup contre la crainte que j'avois de n'avoir pas assés bien rêussi. Je comptois pouvoir vous les faire tenir cet ordinaire et c'est même ce qui m'a fait differer jusqu'à aujourdhuy à vous escrire; mais ils ne sont pas encore finis. Une chose seulement me fait de la peine, outre l'embarras que je vous cause, c'est le prix excessif où je vois que les Graveurs de Hollande mettent leur Ouvrages. Il est vrai que les Insectes que j'ai a faire representer ne sont pas la plupart aussi composés que vos Polypes; et j'ai grand soin de ne pas multiplier les figures pour ne pas faire renchérir un Ouvrage peu considerable en soi. En voici le Plan en deux-mots, sur lequel je vous serai obligé de me dire votre sentiment.

Les Chenilles et Papillons, les Mouches et les Vers desquels elles proviennent, les Pucerons, les Vers, qui se multiplient de boutures, le Toenia, doivent y paraître dans cet Ordre que j'ai préfacé à celui des Changemens de forme, parce qu'il est plus interessant, conduisant le Lecteur de moins singulier au plus singulier. Je fais précéder chaque article d'une Introduction qui contient en raccourci tout ce qui a été decouvert de plus remarquable sur la matiere. Je facilite ainsi l'intelligence de mes propres Observations, et je donne en même tems une idée des Memoires de M. de Reaumur qu'on ne lit pas assés—parce qu'ils sont trop étendus. Enfin je retrace les principes de l'Histoire des Insectes en établissant differens ordres de Transformations. À ces Introductions particulières j'avois d'abord eû dessein d'en joindre une Generale qui auroit renfermé en abregé toute la science des Insectes. Mais j'ai renvoyé l'execution de ce projet au tems ou M. de Reaumur aura achevé son Histoire. Je serai plus en etat alors d'accuser juste sur tout.

Voilà, mon Cher Cousin, tout le plan de mon Receuil. Ce sera un in 4ᵗᵒ de 3 à 400 pages environ. Je n'ai rien encore determiné par raport a l'impression. Oserois je vous demander, à ce sujet comment vous faites avec votre Libraire? On me promet d'ici de l'imprimer aussi bien qu'il est possible.

M. Herrenschwandt qui vous avés vû est bien le même dont je vous ai parlé. Son specifique n'opère surement que sur une Espece de Toenia, sur celle que je nomme *à anneaux courts,* et qui est le l.ᵉʳ des Plates. Il est revenu de Paris à Morat sans avoir eu des occasions suffisantes de l'exercer. On lui a escrit de Hollande qu'il ne s'y trouvoit pas des gens travaillés du Solitaire. Cela me surprend, car je pensois que c'etoit le Pays de l'Europe où il y en avoit le plus.

Je vous embrasse de tout mon Coeur, Monsieur et très Cher Cousin, et suis avec la plus grande estime et le devouement le plus parfait entierement à vous.

M. Bazin mécrit qu'il a commencé un Abregé des Memoires de M. de Reamur. Il a deja donné les Mem.[oires] sur les Abeilles. Mais il s'etend plus dans cet Ouvrage que je ne ferai dans le mien.

Charles Bonnet

25. Trembley to Bonnet.   à La Haie le 24. mars 1744.

Monsieur et Cher Cousin

Pour répondre à vôtre derniére lettre, et aux Questions que me fait mon Frère touchant la gravure de vos Planches, je vous dirai, que je ne crois pas que le graveur V.d. Schley ait le tems de les faire. Il faudra donc comme vous le proposés, se servir d'un autre qui soit moins occupé, et qui, peut être, sera meilleur marché. J'en connois un à Leyde qui, je crois sera vôtre affaire. Il vous sera facile de voir de son ouvrage. Il a gravé une partie des Planches du livre de Swammerdam intitulé *Biblia naturae* &c. Ce livre est imprimé in folio à Leyde en 1737. Je ne doute pas que vous ne le trouviés à Genéve. Ce graveur s'apelle J.v.d. Spyk et son nom se trouve au bas de toutes les Planches du Swammerdam qu'il a gravées. S'il est possible de faire faire une partie de vos Planches à van der Schley, et à un prix raisonnable, je pourrai l'emploie conjointement avec van der Spyk.

Au lieu d'envoier toutes vos Planches à la fois, vous feriés mieux d'en remettre d'abord au plutôt une ou deux à mon Frére. On pourra juger par là de l'ouvrage, convenir avec le graveur, et commencer à le faire travailler. C'est certainement le vrai moien d'expédier. Aiés bien soin de marquer exactement la grandeur que doit avoir la Planche.

Je ne sai quel est vôtre libraire, mais il est bon de l'avertir que nos Graveurs ont accoutumé de ne livrer leur ouvrage aux Libraires qu'en recevant l'argent. Je vous ferai savoir le prix qu'on demande afin que vôtre libraire voie s'il lui convient.

Voici comment j'ai fait avec les miens. Ce sont les fréres Verbeek de Leyde. Je leur ai donc promis de leur donner mon manuscript à condition que je serois le maitre de faire graver autant de Planches que je voudrai, et par le meilleur graveur, que je choisirois le papier, le charactére, et le format, qu'ils me donneroient un certain nombre d'exemplaires, et trois ou quatre livres dont j'avois besoin.

J'attends vos ordres, et je suis de tout mon coeur Tout à vous.

A. Trembley

26. Bonnet to Trembley. A Geneve le 26. Juin 1744

Monsieur et très Cher Cousin

Des affaires de Famille très importantes et divers autres Contretems ont été cause que je n'ai pû encore avoir l'honneur de vous remercier de toutes vos nouvelles attentions au Sujet de mon Ouvrage. Quand je n'aurois d'autres preuves de votre amitié, je n'en pourrois douter le moins du monde. Je crois vous avoir escrit que je comptois faire imprimer ici. Mais les Libraires que j'avois choisis se sont excusés depuis sur la difficulté de l'entreprise et le grand nombre d'ouvrages qu'ils ont sur les bras. Dans cet embarras ayant eû à escrire à M. de Reaumur je l'ai informé de ma situation. D'abord il m'a trouvé un Libraire qui est celui qui debite tous les Livres qui Sortent de l'Imprimerie Royale et il l'a engagé à me donner 100 Exemplaires et à se charger des frais de la gravûre. Ce Libraire est Durand. J'ai donc fait partir pour Paris la 1ere. Partie de mon Manuscrit afin qu'il commence à faire travailler sous les yeux de M. de Reaumur, au jugement duquel j'ai soumis l'ouvrage entier. Je suis actuellement occupé à mettre en etat la 2e partie que j'espere envoyer incessamment. Voila Mon Cher Cousin où j'en suis: Vous voyés que je n'ai rien à désirer soit pour l'impression soit pour la gravûre dès que M. de Reaumur veut bien en prendre connoissance et à l'egard de ses conditions je n'avois point esperé qu'elles seroient aussi avantageuses.

J'attends de jour à autre l'Histoire des Polypes. Je compte que nous l'aurons au moins avant la fin du Mois prochain. Vous pouriés nous en faire parvenir promptement un Exemplaire ou deux par M. de Reaumur. J'ai repeté l'experience de M. Lyonet sur les oeufs des Gros Pucerons du Chêne.

Je suis avec la plus grande Estime et le plus parfait devouement, Monsieur et très Cher Cousin, entierement à vous.

C. Bonnet.

27. Trembley to Bonnet.   à Sorgvliet le 29ᵉ Juin 1744.

Monsieur

Cette lettre vous sera remise par mon Frère avec un exemplaire de mes Mémoires. Il vous aura fait part, j'espére, de ce que je lui ai écrit touchant l'impression de vôtre ouvrage. La guerre continue à déranger la Librairie, et c'est ce qui me fait croire, qu'il ne sera peut-être pas si facile de trouver une bonne et prompte occasion de faire imprimer vôtre ouvrage. Cependant je me persuade que vous n'en trouverés nulle part de meilleure et de plus prompte qu'ici: c'est pourquoi je persiste à vous conseiller d'envoier vôtre manuscript à bon compte dont j'aurai bien soin. Je suis de tout mon coeur et avec un parfait dévoüement

    Monsieur

                    Vôtre très humble et très obéissant
                    Serviteur
                    A. Trembley

28. Trembley to Bonnet.   à Sorgvliet le 31ᵉ Juillet 1744

Monsieur et très Cher Cousin

Vous pouvés être persuadé que si j'avois été chargé de l'Edition de vôtre Ouvrage, je n'aurois rien négligé pour vous témoigner le désir que j'ai de vous servir. Je vous félicite de ce que vous avés déja trouvé un Libraire. Il est certainement plus commode pour vous de faire imprimer à Paris qu'en

Hollande, parce que vous pouvés facilement et sans frais envoier vos Manuscripts par le moien de Mons.ᵣ de Reaumur. Je Souhaite de tout mon coeur que Durand ne vous fasse pas trop attendre, et qu'il ne néglige rien pour les gravures et pour l'impression. Mons.ᵣ de Reaumur m'a écrit il y a quelques jours que ce Libraire se disposoit à imprimer mes Mémoires sur les Polypes, et qu'il comptoit que son Edition seroit plus belle que celle de Hollande, au moins pour l'Impression. Cela me prouve qu'il peut donner de belles Editions, car la mienne est belle. Mes Libraires n'ont rien négligé, et c'est pourquoi, je vois avec peine qu'on se prépare à contrefaire leur ouvrage, parce que cela pourroit leur faire tort. Je ne sai si Durand pourroit en même tems entreprendre vôtre Ouvrage et le mien. Ce seroit une entreprise considerable, vu le grand nombre de gravures qu'il y a. Je crains qu'il ne commence par le mien, et que cela ne nous prive du plaisir de lire le vôtre aussitôt que je le souhaite.

Je compte que mon Frére vous remettra bientôt un exemplaire de mes Mémoires sur les Polypes. Est-il possible qu'on ne puisse pas trouver ces Animaux dans vos quartiers? Ils sont si communs en plusieurs endroits.

J'observe actuellement des Animaux extrémement petits. Ils me paroissent curieux, mais j'aurai besoin de bien de tems pour faire quelques progrès dans leur histoire, si tant est que j'en puisse venir à bout.

Je suis avec une parfaite estime et un entier dévoüement Monsieur et très Cher Cousin. Tout à vous.

A. Trembley.

Voulés vous bien avoir la bonté d'envoier l'incluse à mon Frére le Ministre, pour qu'il la remette à M.ᵣ Baulacre.

29. Trembley to Bonnet. à Sorgvliet le 21ᵉ Aoust 1744

Monsieur et Cher Cousin

Je n'ai pas perdu un instant pour exécuter la commission que vous m'avés donnée dans vôtre lettre du 7.[1] Je l'ai reçue mardi à midi et j'ai envoié le mémoire à M.ᵣ van Swieten[2] le

---

[1]This letter is missing.

[2]Gerhard van Swieten (1700–1772) was a doctor of repute in Holland. Bonnet consulted van Swieten about his problems with his eyes. See Jacques Marx, *Charles Bonnet contre les lumières*, 409–10.

soir par la poste. Vous trouverés ici sa réponse pour la quelle
j'ai donné une guinée. On ne pouvoit pas mieux s'adresser
qu'à M.ʳ van Swieten. C'est un des plus fameux disciples de
Boerhave. Comme les mesures d'un *pot, demi pot,* et *picholettel*
auroient pu lui être inconnues, j'ai expliqué ces termes par
d'autres. Je crois devoir faire une Remarque sur ce que dit
M.ʳ van Swieten, *Victus sit ex hordeo, avena, oryza, fagopyro.* On
ne mange guéres à Geneve l'orge et le ris, qu'en Soupe, au
lieu qu'ici on cuit l'un et l'autre dans l'eau avec un peu de
sel. C'est un ragout très simple et très bon. Voici comment
on le fait. Après que l'orge, par exemple, est cuit, on le separe
de l'eau, on le met dans un plat, et l'on jette dessus un peu
de beurre fondu. Si l'on veut on y ajoute du Sucre. M.ʳ van
Swieten a certainement entendu que le malade usat de ce
mets, que l'on prescrit beaucoup ici. A l'égard du bléd sarasin,
on fait divers gateaux avec sa farine, que M.ʳ Boerhave or-
donnoit pour régime en différentes occasions.

Je souhaite de tout mon coeur que vôtre malade se réta-
blisse.

Je compte que vous avés reçu à présent mes Mémoires sur
les Polypes; et une lettre que je vous ai écrite le 4 de ce mois.

Je sai que Durand veut réimprimer mon ouvrage. J'aime-
rois mieux qu'il ne le fit pas.

Adieu, Mon Cher Cousin. Je suis tout à vous de tout mon
coeur.

A. Trembley

Appendix B

The Bonnet-Cramer Correspondence

In addition to the Bonnet–Trembley letters, to add another dimension to my analysis of the exchange of ideas among the Genevans in response to Trembley's discovery in the 1740s, I have examined the correspondence between Bonnet and Gabriel Cramer, his teacher at the Academy of Calvin. These letters comprise Ms Bonnet 43 in the Public Library of Geneva, *Lettres en original de Mr. Cramer, Professeur de Philosophie* (1740–1749), and Ms Suppl. 384. There are two additional letters from Bonnet to Cramer in the *Dossier ouvert, autographe Bonnet*. I am deeply indebted to Philippe Monnier for his assistance in compiling this appendix of letters. There are a total of 37 letters with six additional essays listed below.

In the present study I have focused on the letters prior to 1747, since these letters have received relatively less scholarly attention. They add a valuable dimension to the debate over the polyp. It appears that parts of only two of these letters were quoted by Maurice Trembley in the footnotes of the *Correspondance inédite entre Réaumur et Abraham Trembley* (pp. 60–61).

A difficulty that will become apparent when the attached list of letters is consulted is the lack of consistent dating of Cramer's letters to Bonnet. However, with a knowledge of the issues discussed, and possessing Bonnet's responses, it is possible to assign tentative dates to the letters which appear to be out of order in the collection. Bonnet's letter to Cramer 23 July 1745 can be found in Ms 43. It appears to be a shorthand copy of the original letter, probably made by Bonnet's secretary.

Of the four sets of correspondence, this is the most philosophical, since in these letters Bonnet's aim is not to report detailed observations, but to attempt, with Cramer's assistance, to make some sense of them. The issue of animal soul, in particular, is brought into focus. However, this set of letters is the most difficult to reconstruct with respect to the flow of ideas since many letters appear to be missing. Because Bonnet often saw his professor in person, some of the issues were most certainly discussed face to face and did not need to be consigned to paper. The examination of the Bonnet–Cramer correspondence reinforces my opinion that Trembley's discovery of the polyp and Bonnet's study of the regeneration of worms, had a determining influence on Bonnet's intellectual formation. Bonnet's studies of parthenogenesis are hardly mentioned. The Bonnet–Cramer letters written between 1747 and 1749 (the year of Cramer's untimely death) center on the problem of free will, one of the great philosophical issues of the eighteenth century. Long excerpts from these letters were published in Bonnet's *Mémoires autobiographiques* (pp. 93–99). They are discussed by Lorin Anderson in *Charles Bonnet and the Order of the Known* (pp. 11–16), and by Jacques Marx in *Charles Bonnet contre les lumières* (pp. 125–30). Despite these perceptive analyses, a complete understanding of the philosophical and biological context in which the Bonnet–Cramer discussions occur must await publication of the full correspondence and the unpublished essays written in response to issues raised by Cuentz.

List of Letters Between Bonnet and Cramer

Bonnet to Cramer *(Ms Suppl. 384)*	Cramer to Bonnet *(Ms Bonnet 43)* [these letters are not bound in chronological order]
	20 December 1740 (Letter 1)
	Letter without date (Letter 2)
	June 1741 (Letter 3)
29 June 1741	
1 August 1741	
18 August 1741	
	19 August 1741 (Letter 5)
7 March 1742	
	Letter without date (Letter 6)
11 June 1742	
17 August 1742	
	Letter without exact date, 1742 (Letter 9)

25 August [1742]	
21 December 1742	
	Letter without date (Letter 4)
	[Response to Bonnet's
	21 December 1742?]
5 June 1743	
22 July 1743	
	Letter without date (Letter 7)
	[Response to Bonnet's
	22 July 1743?]
29 July 1743 (D.o.	
autogr. Bonnet)	
7 August 1743	
14 August 1743	
4 December 1743	
19 July 1745 (D.o.	
autogr. Bonnet)	
	20 July 1745 (Letter 8)
23 July 1745 [draft by	
Bonnet, Ms Bonnet	
43]	
30 June 1747	
31 July 1747	
	11 July 1747 (Letter 10)
Bonnet letters to	
Cramer after	
July 1747 in Ms	
Bonnet 43	
	9 August 1747 (Letter 11)
30 September 1747	
	6 October 1747 (Letter 12)
20 October 1747	
	25 November 1747 (Letter 13)
January 1748	
	25 January 1748 (Letter 14)
	1 July 1749 (Letter 15)

Additional Manuscript Material in Ms Bonnet 43

I. Objections en original de Mr. Cramer contre l'Essai sur
la Liberté du N° V.

II. Réflexions d'Eudoxe sur le premier Tome du Livre in-
titulé *Essai d'un Système nouveau concernant la nature des
Etres Spirituels,* publié par Mr. Cuentz, Conseiller de la
Ville de St. Gal.

III. Sentimens d'Eudoxe sur divers Sujets de Métaphysique,
de Physique, &c.

IV. Portrait de Mr. Arlaud célèbre Peintre de Genève par
Mr. Cramer.

V. Essai sur la Liberté, par C. B. [Charles Bonnet].

VI. Examen de diverses Questions sur la Liberté, par C. B.

Appendix C

The Bonnet–Réaumur, Trembley–Réaumur Letters

The letters which Réaumur exchanged with his Genevan correspondents, Bonnet and Trembley, were scientific in nature. The early letters focus on insects, at first caterpillars and moths, since these were the insects discussed by Réaumur in the early volumes of his *Histoire des insectes*. These early letters were probably typical of other letters which Réaumur answered from naturalists all over Western Europe who were eager to contribute their own observations to Réaumur's already substantial research. Réaumur's early replies are encouraging while the letters of Bonnet and Trembley are full of deference for the influential and esteemed academician. However, once Bonnet and Trembley had made their own significant discoveries, Réaumur's attitude became more collaborative in tone, though he never completely lost his fatherly concern for his much younger correspondents. For example, in the letters to Bonnet he worried about the threatened loss of Bonnet's eyesight, seeking the advice of the most renowned Parisian eye doctor, the octogenarian Claude Deshayes Gendron (1663–1750).

The correspondence between Réaumur and Trembley has been published in the *Correspondance inédite entre Réaumur et Abraham Trembley* (Geneva: Georg, 1943). This painstakingly annotated edition by Maurice Trembley, who inherited the letters as part of the Trembley Family Archives, contains all of the known letters from Réaumur to Trembley. The originals of these letters are now in the archives of Professor George Trembley of Toronto. Abraham Trembley's responses to Réaumur's letters, with the exception of three letters of 27 August, 22 October 1744 and 18 February 1745 in the Archives of the Academy of Sciences are published in the same work. In this published correspondence, which I

have used extensively in my discussion of the discovery of
the polyp in Chapter 4, there are twenty-four letters from
Trembley to Réaumur; sixty-eight letters from Réaumur to
Trembley. Letters written by Trembley to Réaumur after
December 1746 are missing.

All of Trembley's letters beginning with his second letter
are primarily concerned with his research on the polyp.
Though Réaumur raised questions and included his own
observations on the polyp, Trembley's ingenuity in devising
the experiments was clearly his own; Réaumur never suggested a particular experiment to Trembley, spent paragraphs on exclamations and congratulations for their detail
and surprising results. Trembley's notes which formed the
basis of these letters, found in the George Trembley Archives,
also support the view that his observations, while inspired by
Réaumur's example as an exacting observer, were original
and owed little to Réaumur's direct suggestions. For his part,
Réaumur kept Trembley informed of the discoveries of Jean
Etienne Guettard (1715–86), Bernard de Jussieu and Bonnet, who were studying the regeneration of worms and marine animals as a result of Trembley's discovery of regeneration in the polyp. In letter after letter he urged him to
publish. Réaumur frequently asked Trembley to work on salt
water polyps (octopus, squid) which he correctly surmised
shared similar characteristics with fresh water polyps.

After 1745, when Trembley was no longer engaged in
research on the polyp, the tone of the letters shifted and
Réaumur began to share with Trembley more of his own
research which centered first on the artificial hatching of
eggs, the distribution to Dutch ships of his essay on the preservation of bird specimens, and finally his collaboration with
Abbé Joseph-Adrien Lelarge de Lignac (1710?–62) in the
research which formed the basis of de Lignac's *Lettres à un
Amériquain sur l'histoire naturelle, générale et particulière de monsieur de Buffon* (1751), an anonymous attack on Buffon and
his microscopic observations made in collaboration with John
Turberville Needham.

Unfortunately, in contrast to the Réaumur-Trembley correspondence, Bonnet's published correspondence with Réaumur is much less complete. Only the letters of 30 November

1741 and 3 March 1755 from Bonnet have been published in full by Jean Torlais in *Morceaux choisis* (Paris: Gallimard, 1939). In addition, Torlais published a list of the letters in manuscript from Bonnet in his "Inventaire de la correspondance et des papiers de Réaumur conservés aux Archives de l'Académie des Sciences de Paris" in *La Vie et l'oeuvre de Réaumur* (Paris: Presses Universitaires de France, 1962), 18. Pierre Speziali published a list of the Réaumur letters to Charles Bonnet to be found in the archives of the Public and University Library of Geneva in "Réaumur et les savants genevois" in the *Revue d'Histoire des Sciences*, 11 (1958): 70–71. His article was reprinted in *La vie et l'oeuvre de Réaumur*, cited above. This correspondence forms the basis of Torlais's "Un Maître et un élève. Réaumur et Charles Bonnet (d'après leur correspondance inédite)," published as two articles in *Gazette Hebdomadaire des Sciences Médicales de Bordeaux*, 9 October, 16 October 1932, 641–55 and 657–59. Bonnet himself reproduced in his *Mémoires autobiographiques*, ed. Raymond Savioz (Paris: Vrin, 1948), parts of seventeen of the ninety-three letters which he received from Réaumur. Bonnet collected these letters in a manuscript volume preserved at the Library of Geneva: *Lettres en original de M. de Réaumur à Charles Bonnet, 1738–1757* (Ms Bonnet 42).

Of the surviving Bonnet letters in the Paris Academy of Sciences, all twenty-three are unpublished, though large portions appear to have been incorporated into Bonnet's *Traité d'insectologie*. My microfilm does not include several items listed by Jean Torlais: *Essai d'une division générale des insectes* (no date); *Dissertation sur le ver nommé en latin Toenia*, à Genève le 17 février 1748 (121 pp., 1 plate); letter of 9 December 1747; letter of 6 April 1743. I did not have the first edition of this work, but relied on the revised edition, volume one of Bonnet's *Oeuvres d'histoire naturelle et de philosophie* (Neuchâtel: S. Fauche, 1779). The first edition included Bonnet's observations on plant lice which established parthenogenesis and his studies of the regeneration of worms. The other observations which Bonnet made in these early years and which he reported in his letters to Réaumur were not published until the revised edition as a separate section.

Bonnet's letters often have notes in Réaumur's hand on

the first page. They include Bonnet's studies of "liveried" caterpillars, caterpillars which make pendant nests, construction of cocoons with bits of paper, caterpillars which make their nests in the interior of thistles, two new species of antlions. Bonnet reported in detail his studies of the respiration of caterpillars through their stigmata. His research supported Malpighi's view rather than that of Réaumur. In addition, Bonnet concentrated on observations of a "nipple" in various species of caterpillars which Réaumur had regarded as a spinneret. The discovery of this "nipple" was later claimed by de Geer, though Bonnet believed that priority belonged to him.

Réaumur encouraged these early empirical investigations, and was dismayed to discover Bonnet's new speculative concerns after 1748. Referring to Bonnet's threatened loss of eyesight in the wake of his overtaxing observations of plant lice, Réaumur wrote:

I cannot but give great praise to the resignation with which you have borne your state and the manner by which you have softened it with meditation. But it seems that you go too far when you seem to prefer it to what you could do with your eyes which only strain to discover the marvels which the Great Master has effected (*Mém. aut.*, 105).

Though only twenty-three of Bonnet's letters can be found in the Paris Academy of Sciences, their length is excessive. The microfilm of these letters provided me with over three hundred pages of text. I have made complete translations of the first thirteen of these letters and have consulted all the remaining known letters exchanged between them. It is to be regretted that these translations could not be included in the present work, though I would be glad to share them with any scholar wishing to use them. According to Bonnet, his letters were generally copied from a journal of observations which he carefully kept of his insect studies. This journal was not found in the George Trembley Archives; nor in the Public Library of Geneva. Bonnet's day books or journals from 1754 to 1784 are in the George Trembley Archives in eleven volumes. These appear to have been bought by Maurice Trembley from the bookseller Baumgartner of Geneva. They contain excerpts of letters which he considered important. Some of Réaumur's earlier letters are quoted. In addition, there exist drafts of seven of Bonnet's later letters to Réaumur in the archives of the Geneva Public and University Library (Ms Bonnet 70).

A question which should be addressed is why the set of Réaumur letters appears to be complete, while few letters from Bonnet's side of the correspondence have survived. The original letters from Bonnet to Réaumur are fairly complete up to 6 April 1743. Even in these early letters several are missing, including the letters of 11 March 1741 and 19 April 1741 (referred to by Réaumur) which may be of considerable historical value since they were written after the discovery of the polyp, a turning point in Bonnet's intellectual development.

What has become of the Bonnet letters? This question was asked by Bonnet himself with considerable anguish. Bonnet had heard rumors that Réaumur's papers were being sold on the streets of Paris after his death. Bonnet claimed, and it seems reasonable to assume that this is true, that he wrote over a hundred letters to Réaumur. One can judge from Réaumur's responses that many of these letters, particularly those written after 1749, were critical of Buffon. Clearly, Bonnet believed that Buffon was responsible for this loss which he called a theft. In a letter to Albrecht von Haller (1708–1777) he wrote:

. . . Concerning Buffon, I spoke to you previously about the Vol. of manuscripts of M. de Réaumur, which included a great number of my letters. I wished to verify the fact and I learned by a very sure source that the Academy, heir of these manuscripts, having named two friends of his as commissioners to receive this bequest, Buffon went to the king and had him execute an order to put himself alone in charge. *After that all was up for grabs.* These were the very terms of the person who wrote to me . . . If Buffon took the trouble to read these letters, he would have seen things about which he would not have been pleased. You know his hatred for our worthy friend and how much he mistreated him in several places in his writings. Once M. de Réaumur wished to have me publish a letter which I wrote to him on the Ideas of his adversary. I answered him immediately that this letter was only for him and that I preferred peace to everything. (Letter of 23 May 1760. Original text in Otto Sonntag, ed., *The Correspondence between Albrecht von Haller and Charles Bonnet,* 203.)

There is no indication given by Torlais of where the letters he cites can be located. A copy of Bonnet's letter of 18 October 1760 to Duhamel about Buffon's misdeeds with respect to Réaumur's letters was furnished to me courtesy of the American Philosophical Society, Duhamel Papers, 580: D 881, no. 58, 1. Some information about Réaumur's differences with Buffon and his "clique" can be found in J.L. Heilbron, *Electricity in the 17th and 18th Centuries* (Berkeley: University of California Press, 1979), 346–48. It is hoped that some of these early letters from Bonnet will turn up in private collections.

List of Letters between Bonnet and Réaumur

Bonnet Letters (Paris Academy of Sciences)	Réaumur Letters (Public Library of Geneva)
4 July 1738 (7 pp.)	
	22 July 1738[1]
3 October 1738 (8 pp.) [wrong date on 1st page]	
	16 December 1738
22 June 1739 (12 pp.)	
	10 July 1739
27 July 1739 (24 pp.)	
29 July 1739 (2 pp.)	
	11 August 1739
19 August 1739 (8 pp.)	
	23 November 1739
3 December 1739 (20 pp.)	
	20 January 1740
17 February 1740 (8 pp.)	
	4 June 1740
13 July 1740 (24 pp.)	
	5 August 1740[2]
	5 September 1740
[4 November 1740 missing] 23 November 1740 (16 pp.)	

	4 December 1740
30 December 1740 (10 pp.)	
	13 January 1741
[11 March 1741 missing]	
[19 April 1741 missing]	
	30 April 1741
	June, 1741[2]
[28 July 1741 missing]	
	7 August 1741[1]
25 August 1741 (31 pp.)	
	5 September 1741
4 November 1741 (8 pp.)	
	30 November 1741[3]
3 February 1742 (44 pp.)	
	28 February 1742[2]
7 March 1742 (12 pp.)	
[Mémoire, not letter]	
	23 May 1742
17 June 1742 (?) (12 pp.)	
[last page missing; listed	
in Torlais as 1752]	
23 June 1742 (30 pp.) [mis-	
dated on 1st page]	
	8 August 1742[1]
15 August 1742 (22 pp.)	
[not 20 as in Torlais]	
5 November 1742 (42 pp.)	
	11 November 1742
	21 December 1742
7 January 1743 (43 pp.)	
[6 April 1743 listed in	
Torlais, but not in my mi-	
crofilm]	
	12 April 1743
	25 August 1743
[30 August 1743 missing]	
[18 September 1743	
missing]	

10 November 1743
18 December 1743[1]
15 April 1744
15 May 1744
13 July 1744
30 August 1744
 2 January 1745
19 February 1745
23 March 1745
27 June 1745
 5 September 1745
 4 December 1745
23 December 1745
 4 February 1746
14 April 1746
15 June 1746

3 August 1746 (8 pp.)
5 August 1746 (4 pp.)

 5 February 1747[1]
18 February 1747
 5 June 1747
17 August 1747[1]
22 November 1747

9 December 1747 [listed in
Torlais, but not on my
microfilm]

15 January 1748

17 February 1748 [*Mémoire*
listed by Torlais, but not
on my microfilm]

 9 July 1748
 6 September 1748
15 December 1748
15 January 1749
 1 September 1749
18 December 1749[1]
30 April 1750
27 June 1750
26 July 1750

	27 July 1750
	31 August 1750[1]
	30 November 1750
	9 February 1750
	27 April 1750
	26 June 1751
June, 1751 (draft)	
	7 August 1751
9 August 1751 (draft)	
	29 August 1751[1]
27 November 1751 (draft)	
	10 December 1751
26 June 1752 (draft)	
	7 July 1752
26 July 1752 (draft)	
	2 August 1752
November, 1752 (draft)	
	15 December 1752
	9 May 1753
23 May 1753 (draft)	
	9 July 1753
	31 August 1753
	27 January 1754
	9 February 1754[1]
	14 February 1754
	24 February 1754
	14 March 1754
	29 June 1754
	25 August 1754
	27 November 1754
	3 January 1755
	3 March 1755[3]
	20 May 1755
	28 May 1755
	30 July 1755
	2 October 1755
	11 December 1755
	24 January 1756

1 March 1756
5 April 1756[1]
9 April 1756
22 August 1756
6 November 1756
29 November 1756
6 March 1757

(Réaumur died 18 October   7 August 1757
   1757.)

---

[1]Fragment published in *Mémoires autobiographiques de Charles Bonnet,* ed. Raymond Savioz (Paris: Vrin, 1948).

[2]Fragment published in *Correspondance inédite entre Réaumur et Abraham Trembley,* ed. Maurice Trembley (Geneva: Georg, 1943).

[3]Published in full by Jean Torlais, *Morceaux choisis* (Paris: Gallimard, 1939).

Selected Bibliography

PRIMARY SOURCES

Manuscript Collections
Geneva. Bibliothèque Publique et Universitaire
  Ms Bonnet 1
  Ms Bonnet 24
  Ms Bonnet 42
  Ms Bonnet 43
  Ms Bonnet 70
  Ms Suppl. 384
  Ms S.H. 242
  Ms Collection Condet Ms Suppl. 363
  Dossier ouvert, autographe Trembley

Paris. Archives de l'Académie des Sciences
  Papers of Réaumur. Bonnet (DB) Letters from Bonnet to
    Réaumur

Philadelphia. American Philosophical Society
  Duhamel Papers. 580: D 881, no. 58, 1–2

Toronto. George Trembley Archives
  Carton of letters from Bonnet to Trembley
  Carton of letters from Allamand to Trembley
  Bound Day Book in Trembley's hand containing religious
    reflections
  Bound ms volume: *Cours de Logique* par feu M. Cramer,
    Professeur de Philosophie, etc. l'Acad. de Genève re-
    cueilli par Ch. Bonnet

London, England. The British Library
  Egerton Ms 1726 Letter from Trembley to William Ben-
    tinck

Books and Papers
Baker, Henry. *An Attempt Towards a Natural History of the Pol-*

ype: In a Letter to Martin Folkes, Esq., President of the Royal Society. London: Dodsley, 1743.

Boerhaave, Hermann. *Elements of Chemistry.* Translated by Timothy Dallowe. London: J. Clarke, A. Millar and J. Gray for J. and J. Pemberton, 1735.

Bonnet, Charles. *Mémoires autobiographiques de Charles Bonnet de Genève.* Edited by Raymond Savioz. Paris: Vrin, 1948.

————. "Observations sur quelques auteurs d'histoire naturelle." In *Correspondance littéraire, philosophique et critique par Grimm, Diderot, Raynal, Meister, etc.* Edited by Maurice Tourneux. Vol. 4, pp. 163–71. Paris: Garnier Frères, 1878.

————. *Oeuvres d'histoire naturelle et de philosophie, 1779–1783.* 8 vols. Neuchâtel: S. Fauche, 1779–83. Vol. 1: *Traité d'insectologie.* 1779.

————. "Observations on Insects." In *Philosophical Transactions,* no. 470 (1743), 458–87.

Bourguet, Louis. *Lettres philosophiques.* 2nd ed. Amsterdam: Marc-Michel Rey, 1762.

Butler, Joseph. *The Analogy of Religion.* Cambridge: Hilliard and Brown, 1827.

Clarke, Samuel. *A Demonstration of the Being and Attributes of God: More Particularly in Answer to Mr. Hobbs, Spinoza and their Followers, Wherein the Notion of Liberty is Stated, and the Possibility and the Certainty of it Proved, in Opposition to Necessity and Fate.* London: Botham, 1704. Reprint ed., Stuttgart-Bad Cannstatt: Frommann, 1964.

Derham, William. *Physico-Theology: Or, a Demonstration of the Being and Attributes of God from His Works of Creation. Being the Substance of Sixteen Sermons Preached in St. Mary-le-Bow Church, London, at The Honorable Mr. Boyle's Lectures, in the Years 1711, and 1712. With Large Notes and Many Curious Observations.* 2nd ed. London: W. Innys, 1714.

Dessaussure [sic], H[orace] B[énédict]. *Eloge historique de Charles Bonnet.* [Pamphlet, 1793].

"Diverses observations de physique et d'histoire naturelle." *Histoire de l'Académie Royale des Sciences de l'année 1741 avec les mémoires de mathématiques & de physique pour la même année.* Amsterdam: Pierre Mortier, 1747. Pp. 44–48.

Fontenelle, Bernard. *Entretiens sur la pluralité des mondes*

(1687). Edited by Robert Shackelton. Reprint ed. Oxford: Clarendon Press, 1955.

Fouchy, Grandjean de. "Eloge de M. Réaumur." *Histoire de l'Acdémie Royale des Sciences de l'année 1757 avec les mémoires de mathématiques & de physique pour la même année.* Amsterdam: J. Schrender, 1769. Vol. 1 of 3. Pp. 306–32.

'sGravesande, G[uillaume] J. *Oeuvres philosophiques et mathématiques de Mr. G.J. 'sGravesande, assemblées & publiées par Jean Nic. Seb. Allamand, qui y a ajouté l'histoire de la vie & des écrits de l'auteur.* 2 vols. Amsterdam: Rey, 1774.

Leeuwenhoek, Anton van. "Part of a Letter from Mr. Antony van Leeuwenhoek, F.R.S. Concerning Green Weeds Growing in the Water, and Some *Animalcula* Found About Them." *Philosophical Transactions.* No. 23 (1703–4): 1304–11.

Lesser, Wilhelm Friedrich. *Theologie des insectes ou démonstration des perfections de Dieu dans tout ce qui concerne les insectes, traduit de l'Allemand de M. Lesser avec des remarques de M. P. Lyonnet,* 2 vols. Paris: Hagues-Daniel Chaubert & Laurant Durand, 1745.

Pardies, Ignace-Gaston. *Discours de la connoissance des bestes, 1672.* Edited by Leonora Cohen Rosenfield. New York: Johnson Repr., 1972.

Pluche, Abbé [Noël Antoine]. *Spectacle de la nature or Nature Display'd, Being Discourses on Such Particulars of Natural History as Were Thought Most Proper to Excite the Curiosity and Minds of Youth.* 10th ed. London: Davis and Reymers, 1766.

Réaumur, René-Antoine Ferchault de. *Mémoires pour servir à l'histoire des insectes.* 6 vols. Paris: Imprimerie Royale, 1734–42.

———. "Sur les diverses reproductions qui se font dans les écrevisses, les omars, les crabes, &c. entr'autres sur celles de leurs jambes & de leurs écailles." *Histoire de l'Académie Royale des Sciences de l'année 1712 avec les mémoires de mathématiques & de physique pour la même année.* Paris: Pankoucke, 1777. Pp. 295–321.

Trembley, Abraham. *Mémoires, pour servir à l'histoire d'un genre de polypes d'eau douce, à bras en forme de cornes.* Leide: Jean & Herman Verbeek, 1744.

[Trembley, Jean.] *Mémoire historique sur la vie et les écrits de*

Monsieur Abraham Trembley. Neuchâtel: S. Fauche, 1787.
Microfilm edition, National Library of Medicine.
Trembley, Maurice, ed. *Correspondance inédite entre Réaumur
et Abraham Trembley.* Genève: Georg, 1943.

SECONDARY SOURCES

Unpublished Books and Articles
Dahm, John. "Science and Religion in Eighteenth-Century
England: The Early Boyle Lectures and the Bridgewater
Treatises." Ph.D. Dissertation. Case Western Reserve Uni-
versity. 1969.
Montandon, Cleopatra. "The Development of Science in Ge-
neva in the XVIIIth and XIXth Centuries: The Case of a
Scientific Community." Ph.D. Dissertation. Columbia Uni-
versity. 1973.
Trembley, Maurice. "Un Philosophe du XVIIIe siècle:
Charles Bonnet." June, 1895. (Typewritten.)
————. "Réaumur et ses correspondants Genevois." Confér-
ence faite à l'aula de l'Université, Geneva, December 1902.
(Typewritten.)
————. "Charles Bonnet et Abraham Trembley—leurs re-
lations d'amitié." Conférence faite à l'aula de l'Université,
Genèva, 19 December 1902. (Typewritten.)

General References
Dictionary of Scientific Biography. Charles Gillispie, ed. New
York: Scribner's, 1973–78.

Books and Articles
Anderson, Lorin. *Charles Bonnet and the Order of the Known.*
(Studies in the History of Modern Science, vol. 11. Boston:
Reidel, 1982.
————. "Charles Bonnet's Taxonomy and Chain of Being,"
Journal of the History of Ideas 37 (1976): 45–58.
Armstrong, Brian. *Calvinism and the Amyraut Heresy.* Madison:
University of Wisconsin Press, 1969.
Baker, John R. *Abraham Trembley of Geneva.* London: Arnold,
1952.

Balz, Albert G.A. "Cartesian Doctrine and the Animal Soul—An Incident in the Formation of Modern Philosophical Tradition." In *Studies in the History of Ideas* 3: 117–77. New York: Columbia University Press, 1935.

Barber, William H. *Leibniz in France from Arnauld to Voltaire*. Oxford: Clarendon Press, 1952.

Bizer, Ernst. "Reformed Orthodoxy and Cartesianism." Translated by Chalmers MacCormick. In *Translating Theology into the Modern Age*. Edited by Rudolf Bultmann et al., 2: 20–82. New York: Harper Torchbooks, 1965.

Bodemer, Charles W. "Overtures to Behavioral Science: 18th and 19th Century Ideas Relating to Light and Animal Behavior," *Episthème* 1 (1967): 135–52.

———. "Regeneration and the Decline of Preformationism in Eighteenth Century Embryology," *Bulletin in the History of Medicine* 38 (1964): 20–31.

Borgeaud, Charles. *Histoire de l'Université de Genève*. Vol. 1: *L'Académie de Calvin (1559–1798)*. Genève: Georg, 1900.

Bouillier, Francisque. *Histoire de la philosophie cartésienne*. 3rd ed. Paris: Delagrave, 1868.

Bowler, Peter. "Preformation and Pre-Existence in the Seventeenth Century: A Brief Analysis," *Journal of the History of Biology* 4 (1971): 221–44.

Brunet, Pierre. *L'Introduction des théories de Newton en France au XVIIIᵉ siècle, avant 1738*. Paris: Albert Blanchard, 1931.

———. *Les Physiciens hollandais et la méthode expérimentale en France au XVIIIᵉ siècle*. Paris: Albert Blanchard, 1926.

Budé, Eugène de. *Vie de Jean-Robert Chouet*. Geneva: Raymond, 1899.

Caraman, R. de. *Charles Bonnet philosophe et naturaliste, sa vie et ses oeuvres*. Paris: A. Vaton, 1859.

Centre International de Synthèse. *La Vie et l'oeuvre de Réaumur 1683–1757*. Paris: Presses Universitaires de France, 1962.

Cole, Francis J. *Early Theories of Sexual Generation*. Oxford: Clarendon Press, 1930.

———. "The 'Biblia naturae' of Swammerdam," *Nature* 165 (195):511.

———. "Jan Swammerdam 1637–80," *Nature* 139 (1937):218–20; 287.

Corcos, Alain F. "Fontenelle and the Problem of Generation

in the Eighteenth Century," *Journal of the History of Biology* 4 (1971):363–72.

Cragg, Gerald R. *Reason and Authority in the Eighteenth Century.* London: Cambridge University Press, 1963.

Delaporte, François. "Des Organismes problématiques," *Dix-huitième Siècle* 9 (1977):49–59.

———. *Nature's Second Kingdom.* Translated by A. Goldhammer. Cambridge, Mass.: MIT Press, 1981.

Delorme, Susanne. "L'Académie Royale des Sciences: ses correspondants en Suisse," *Revue d'Histoire des Sciences et de leurs Applications* 4 (1951):159–70.

Duchesneau, François. "Leibniz et la théorie physiologique," *Journal of the History of Philosophy* 14 (1976):281–300.

Ehrard, H. "Die Entdeckung der Parthenogenesis durch Charles Bonnet," *Gesnerus* 3 (1946):15–27.

Ehrard, Jean. *L'Idée de nature dans la première moitié du XVIIIe siècle.* Vol. 1. Paris: Ecole Pratique des Hautes Etudes, VI section, 1963.

Fontenay, Elizabeth. "La Bête est sans raison," *Critique* (Paris) 34 (1978):707–29.

Gasking, Elizabeth. *Investigations into Generation 1651–1828.* Baltimore: Johns Hopkins University Press, [1967].

———. *The Rise of Experimental Biology.* New York: Random House, 1970.

Geisendorf, Paul-F. *Les Trembley de Genève.* Genève: Julien, 1970.

Glass, Bentley, Owsei Temkin, and William L. Strauss, Jr., eds. *Forerunners of Darwin, 1745–1859.* Baltimore: Johns Hopkins Press, 1959.

Guyénot, Emile. "La Zoologie à Genève et en France," *Revue Economique Franco-Suisse.* June, 1959. Pp. 43–45.

Hahn, Roger. *The Anatomy of a Scientific Institution: The Paris Academy of Sciences, 1666–1803.* Berkeley: University of California Press, 1971.

Hall, Thomas S. *Ideas of Life and Matter.* Vol. 1. Chicago: University of Chicago Press, 1969.

Hastings, Hester. *Man and Beast in French Thought of the Eighteenth Century.* Johns Hopkins Studies in Romance Literature and Languages, vol. 27. Baltimore: Johns Hopkins University Press, 1936.

Heyd, Michael. *Between Orthodoxy and Enlightenment: Jean-Robert Chouet and the Introduction of Cartesian Science in the Academy of Geneva.* The Hague: Nijhoff, 1982.

Hoskyn, F.P. "The Relations of Malebranche and Leibniz on Questions in Cartesian Physics," *The Monist* 40 (1930):131–45.

Irving, J.A. "Leibniz' Theory of Matter," *Philosophy of Science* 3 (1936):208–14.

Jacob, Margaret C. *The Radical Enlightenment: Pantheists, Freemasons and Republicans.* London: George Allen and Unwin, 1981.

Kanaev, I.I. *Hydra: Essays on the Biology of Fresh Water Polyps.* Edited by Howard M. Lenhoff. Translated by T. Burrows and Howard M. Lenhoff from the Russian. Moscow: Soviet Academy of Sciences, 1952. Published in limited edition, University of Miami, 1969.

Kiernan, Colm. "The Enlightenment and Science in Eighteenth Century France." In *Studies on Voltaire and the 18th Century.* Edited by Theodore Besterman. Vol. 59A. Banbury, England: The Voltaire Foundation, 1973.

King, Lester S. *The Growth of Medical Thought.* Chicago: University of Chicago Press, 1963.

Lenhoff, Howard M. "Tissue Grafting in Animals; Its Discovery in 1742 by Abraham Trembley as he Experimented with Hydra," *Biological Bulletin* 177(1984):1–10.

Lenhoff, Howard M., and Pierre Tardent, eds. *From Trembley's Polyps to New Directions in Research on Hydra: Proceedings of a Symposium Honoring Abraham Trembley (1710–1784). Archives des Sciences. (Genève)* 38 (1985).

Lenhoff, Sylvia G., and Howard M. Lenhoff. *Hydra: the Birth of Experimental Biology—1744: Abraham Trembley's Mémoires concerning the Polyps.* Pacific Grove, CA: Boxwood Press, 1986.

Lovejoy, Arthur. *The Great Chain of Being.* Cambridge: Harvard University Press, 1936; Harper Torchbook, 1960.

McNeill, John T. *History and Character of Calvinism.* New York: Oxford University Press, 1954.

Marx, Jacques. "L'Art d'observer au XVIIIᵉ siècle: Jean Senebier et Charles Bonnet," *Janus* 61 (1974): 201–20.

————. "Charles Bonnet et les courantes biophilosophiques en Hollande," *Epistème* 10 (1976):190–208.

————. *Charles Bonnet contre les lumières: 1738–1850.* In *Studies on Voltaire and the 18th Century.* Edited by Theodore Besterman. Vols. 156–57. Banbury, England: The Voltaire Foundation, 1976.

————. "La Préformation du germe dans la philosophie biologique au XVIIIᵉ siècle," *Tijdschrift voor de Studie von de Verlichting* 1 (1973):397–428.

Meyer, Arthur William. *The Rise of Embryology.* Stanford: Stanford University Press, 1939.

Montandon, Cléopâtre. "Sciences et société à Genève aux XVIIIᵉ et XIXᵉ siècles," *Gesnerus* 32 (1975):16–34.

Mossner, Ernest Campbell. *Bishop Butler and the Age of Reason.* New York: Macmillan, 1936.

Müller, Gerhard H. "René-Antoine F. de Réaumur et la classification des insectes," *Rivista di Biologia, Biology Forum* 79 (1986): 203–28.

Nordenskiöld, Eric. *The History of Biology.* New York: Tudor Publishing, 1935.

Raskolnikoff, Mouza. "De l'éducation au Siècle des Lumières: Louis de Beaufort Gouverneur du Prince de Hesse Hombourg d'après des lettres inédites," *Journal des Savants* (Paris) (janvier-mars, 1982):77–93.

Ritterbush, Philip C. *Overtures to Biology.* New Haven: Yale University Press, 1964.

Roe, Shirley A. *Matter, Life and Generation: 18th Century Embryology and the Haller-Wolff Debate.* Cambridge: Cambridge University Press, 1981.

————. "The Development of Albrecht von Haller's Views on Embryology," *Journal of the History of Biology* 8:167–90.

Roger, Jacques. *Les Sciences de la vie dans la pensée française du XVIIIᵉ siècle: la génération des animaux de Descartes à l'Encyclopédie.* Paris: Armand Colin, 1963.

————. "Leibniz et les sciences de la vie," *Akten des internationalen Leibniz Kongresses.* Wiesbaden: 1969. Vol. 2. Pp. 209–19.

Rosenfield, Leonora Cohen. *From Beast-Machine to Man-Machine.* New York: Oxford University Press, 1941.

Rostand, Jean. *La Parthénogenèse animale.* Paris: Presses Universitaires de France, 1950.

Rudolph, Gérard. "Les Débats de la transplantation expérimentale—considérations de Charles Bonnet (1720–93) sur la 'greffe animale,'" *Gesnerus* 34 (1977):50–68.

Savioz, Raymond. "Un Maître et un disciple au XVIIIᵉ siècle (Charles Bonnet et Réaumur)," *Thalès 4* (1940):100–12.

————. *La Philosophie de Charles Bonnet de Genève.* Paris: Vrin, 1948.

Sayous, André. "Charles Bonnet: sa vie et ses travaux d'après une correspondance inédite," *Revue des Deux Mondes* 12 (1855):49–81.

————. *Le Dix-huitième Siècle à l'étranger.* 2 vols. 1861. Reprint (2 vols. in 1). Geneva: Slatkine, 1970.

Schiller, Joseph. "La Notion d'organisation dans l'oeuvre de Louis Bourguet (1678–1742)," *Gesnerus* 32 (1975):87–97.

————. *Physiology and Classification: Historical Relations.* Series Université de Compiègne. Paris: Maloine, 1980.

Schofield, Robert E. "An Evolutionary Taxonomy of Eighteenth Century Newtonians," *Studies in Eighteenth Century Culture* 7 (1978):175–92.

————. *Mechanism and Materialism: British Natural Philosophy in an Age of Reason.* Princeton: Princeton University Press, 1972.

Schopfer, W.H. "L'Histoire des théories relatives à la génération, aux 18ᵉᵐᵉ et 19ᵉᵐᵉ siècles," *Gesnerus* 2 (1945): 81–103.

Schrecker, Paul. "Malebranche et le préformisme biologique," *Revue Internationale de Philosophie* 1 (1938):77–97.

Smith, Cyril Stanley, ed. *Réaumur's Memoirs on Steel and Iron.* 1722. Chicago: University of Chicago Press, 1956.

Speziali, Pierre. "Réaumur et les savants Genevois," *Revue d'Histoire des Sciences* 11 (1958):68–80.

————. "Gabriel Cramer et ses correspondants." Conférence faite au Palais de la Découverte. Paris, 1958.

Torlais, Jean. "Un Maître et un élève. Réaumur et Charles Bonnet (d'après leur correspondance inédite)," *Gazette Hebdomadaire des Sciences Médicales de Bordeaux.* October, 1932. Pp. 641–55; 657–9.

————. *Réaumur: un esprit encyclopédique en dehors de l'Encyclopédie.* Paris: Albert Blanchard, 1961.

Van Seters, W. H. *Pierre Lyonet, 1706–1789.* La Haye: Martinus Nijhoff, 1962.

Vartanian, Aram. "Trembley's Polyp, La Mettrie, and Eighteenth-Century French Materialism," *Journal of the History of Ideas* 11 (1950):259–86.

Webster, Charles. "The Recognition of Plant Sensitivity by English Botanists in the Seventeenth Century," *Isis* 57 (1966):5–23.

Wheeler, Morton, ed. *Réaumur's Natural History of Ants.* New York: Knopf, 1926.

Whitman, Charles O. "Bonnet's Theory of Evolution," and "The Palingenesia and the Germ Doctrine of Bonnet." In *Biological Lectures Delivered at the Marine Biological Laboratory at Woods Holl* [sic] *in the Summer of 1894,* 3:225–72. Boston: Ginn, 1896.

Yolton, John W. *Thinking Matter: Materialism in Eighteenth-Century Britain.* Minneapolis: University of Minnesota Press, 1983.

INDEX